Plant Cell Monographs

Volume 24

Series editor
Peter Nick
Karlsruhe, Germany

More information about this series at http://www.springer.com/series/7089

Vaidurya Pratap Sahi • František Baluška
Editors

The Cytoskeleton

Diverse Roles in a Plant's Life

 Springer

Editors
Vaidurya Pratap Sahi
Botanical Institute
Karlsruhe Institute of Technology
Karlsruhe, Germany

František Baluška
IZMB, Department of Plant Cell Biology
University of Bonn
Bonn, Nordrhein-Westfalen, Germany

ISSN 1861-1370   ISSN 1861-1362   (electronic)
Plant Cell Monographs
ISBN 978-3-030-33530-4   ISBN 978-3-030-33528-1   (eBook)
https://doi.org/10.1007/978-3-030-33528-1

This Springer imprint is published by the registered company Springer Nature Switzerland AG.
The registered company address is: Gewerbestrasse 11, 6330 Cham, Switzerland

Preface

The cytoskeleton, microtubules, and actin filaments together have been found to play diverse roles in the growth and development of plants. The role of cytoskeleton is well established in providing shape and size to plants cells. Be it the jigsaw shape of pavement cells or the conical shape of petal epidermis or the grain shape of rice, it is all brought about by the coordination of the cytoskeleton. The plant cell uses microtubules for trafficking CESA to the cell wall. The orientation of microfibrils corresponds to the microtubular orientation, thereby giving a functional role to the microtubules in context to cell wall structure. The cytoskeleton not only provides for the geometric dimensions but is also very important in physiological processes. Actin filaments are known to play roles in dynamics of stomata and chloroplast, both of which have physiological consequences which are related to adaptations pertaining to abiotic stresses. Recent studies show the interactions between microtubules and actin filaments. Also, it is evident from hormonal cross talks that the cytoskeleton in plants is needed for proper distribution of hormones. Advances in imaging techniques have made the functional studies of cytoskeleton more fascinating in plants. In this book, the authors would like to bring out the role of plant cytoskeleton in context to its interactions and functional affinity to other cellular organelles.

Karlsruhe, Germany   Vaidurya Pratap Sahi
Bonn, Germany   František Baluška

Contents

1 Cortical Region of Diffusively Growing Cells as a Site of Actin–Microtubule Cooperation in Cell Wall Synthesis 1
Kateřina Schwarzerová and Judith García-González

2 Insights into the Cell Wall and Cytoskeletal Regulation by Mechanical Forces in Plants 23
Yang Wang, Ritika Kulshreshtha, and Arun Sampathkumar

3 Chloroplast Actin Filaments Involved in Chloroplast Photorelocation Movements 37
Masamitsu Wada and Sam-Geun Kong

4 Diversity of Plant Actin–Myosin Systems 49
Takeshi Haraguchi, Zhongrui Duan, Masanori Tamanaha, Kohji Ito, and Motoki Tominaga

5 Actin Cytoskeleton and Action Potentials: Forgotten Connections 63
F. Baluška and S. Mancuso

6 The Actomyosin System in Plant Cell Division: Lessons Learned from Microscopy and Pharmacology 85
Einat Sadot and Elison B. Blancaflor

7 Cooperation Between Auxin and Actin During the Process of Plant Polar Growth 101
Jie Liu and Markus Geisler

8 Interactions Between the Plant Endomembranes and the Cytoskeleton 125
Pengfei Cao and Federica Brandizzi

Chapter 1
Cortical Region of Diffusively Growing Cells as a Site of Actin–Microtubule Cooperation in Cell Wall Synthesis

Kateřina Schwarzerová and Judith García-González

Abstract The cortical cytoskeleton, consisting of microtubules and actin filaments, plays a crucial role in the shaping and synthesis of the cell wall. Microtubules and actin filaments engage in cross-talk in plant cells, and this interplay is mediated by a number of molecular interactors and signaling pathways. This work is focused on the interconnected role of actin and microtubules during cell wall formation. Proteins possibly involved in the cross-talk between microtubules and actin filaments for cell wall assembly control are described, and pathways connecting their function in specialized cells with complex shapes are discussed. These include *Arabidopsis* trichomes, interdigitating epidermal pavement cells, and xylem vessel cells. Mutual interactions between microtubules and actin filaments in these cells are based on restrictive cooperation, often controlled by overlapping regulatory pathways, rather than direct cross-link between both cytoskeletons. A specialized formation of the cytoskeletal structure, the preprophase band, is further discussed as an example of direct microtubules and actin filaments cross-linking.

1.1 Cortical Cytoplasm

The cortical cytoskeleton, which consists of microtubules (MTs) and actin filaments (AFs), plays a crucial role in the synthesis and shaping of the cell wall (Szymanski and Cosgrove 2009). The role of MTs in cell wall deposition is under intense investigation, which has uncovered many molecular details of the process. On the other hand, although the importance of AFs in cell wall deposition is unquestionable, the molecular mechanisms underlying its role are less clear. Cortical MTs and AFs clearly interact with each other (for review see Takeuchi et al. 2017). MTs and AFs

K. Schwarzerová (✉) · J. García-González
Department of Experimental Plant Biology, Faculty of Science, Charles University, Prague, Czech Republic
e-mail: katerina.schwarzerova@natur.cuni.cz

© Springer Nature Switzerland AG 2019
V. P. Sahi, F. Baluška (eds.), *The Cytoskeleton*, Plant Cell Monographs 24,
https://doi.org/10.1007/978-3-030-33528-1_1

cross-talk is mediated by a number of molecular interactors (reviewed in Petrášek and Schwarzerová 2009; Krtková et al. 2016; Takeuchi et al. 2017). This work is focused on MTs–AFs cross-talk mechanisms involved in cell wall assembly. Firstly, from the list of proteins that cross-link MTs and AFs, candidate proteins possibly involved in cell wall assembly control are selected. Secondly, we discuss signaling pathways connecting MT and AF cytoskeletons in specialized cells with polar expansion. These include *Arabidopsis* trichomes cells, interdigitating epidermal pavement cells, and xylem vessel cells depositing the secondary cell wall in unique patterns. Current data suggest that in these specialized cells, MTs and AFs delimit different cortical zones on the plasma membrane for localized deposition of the cell wall with distinct mechanical properties. Mutual interactions between MTs and AFs are based on restrictive cooperation, often controlled by one upstream regulator rather than direct cross-link between both types of polymers. A specialized formation of cytoskeletal structure, the preprophase band, is further discussed as an example of direct MTs and AFs cross-linking.

1.2 Plant Cell Wall and Its Deposition

Plant cell walls are semirigid, composite structures. The main component is polysaccharides, which strengthen the cell wall while maintaining enough plasticity for growth (Cosgrove 2005; Szymanski and Cosgrove 2009; Bashline et al. 2014). The polysaccharides that make up the cell wall are cellulose, hemicelluloses, and pectins.

1.2.1 Microtubules and Cellulose Synthesis

Cellulose is the main load-bearing polymer of the cell wall. The molecular mechanism of cellulose synthesis is based on a multimeric enzyme complex, localized in the plasma membrane. Enzymes that synthesize cellulose are called cellulose synthases (CESA). The CESA complexes move within the plane of plasma membrane during cellulose synthesis. Each protein complex consists of 18–36 single catalytic subunits of CESA; 18 subunits are favored by current models (Vandavasi et al. 2016). Cellulose is synthesized into the cell wall, where it crystallizes, and the crystallization process is believed to generate lateral movement of CESA complexes within the plasma membrane (Paredez et al. 2006).

Since the deposition of cellulose significantly influences cell wall properties, plant cells control CESA movement through a unique mechanism. The process involves cortical MTs in close association with the plasma membrane to define the direction of CESA movement (Paredez et al. 2006). Several proteins mediate the interaction between the MTs and the CESA complex. An MT-associated protein (MAP)

CELLULOSE SYNTHASE INTERACTING1 (CSI1) links cortical MTs to CESA complexes (Bringmann et al. 2012; Li et al. 2012). After the identification of CSI1, several other proteins involved in the control of CESA attachment to MTs were discovered. This suggests that cellulose synthesis is controlled by large protein machinery. COMPANION OF CELLULOSE Synthase (CC) proteins (Endler et al. 2016) are also a component of the CESA complex. Here, CC proteins function as a MAP independently of CSI1, and their function is based on the control of microtubular dynamics under stress conditions, ensuring that cellulose synthesis is sustained (Kesten et al. 2019). The physical interaction between MTs and CESA is controlled by another kind of MAP called cellulose synthase–microtubule uncouplings (CMU) (Liu et al. 2016). The cellulose-producing protein factory probably has more components regulating cellulose synthesis, such as KORRIGAN (Vain et al. 2014), CHITINASE-LIKE1 (Sanchez-Rodriguez et al. 2012), or COBRA (Liu et al. 2013). These proteins partake in cellulose synthesis and may be associated with CESA complexes as well.

1.2.2   Actin and Cellulose Synthesis

MTs, CESA protein complexes, and associated regulatory proteins are crucial for cellulose synthesis and plants impaired in this mechanism usually display strong morphogenic defects. However, what is the role of the actin cytoskeleton in cellulose deposition, and is microtubular and actin cytoskeletons cross-talk important for this process? The cortical actin cytoskeleton is spatially and structurally associated with MTs (Sampathkumar et al. 2011). The actin cytoskeleton is involved in the process of cellulose deposition as well. For example, 5-day-old etiolated seedlings of *act2act7* mutants contain less cellulose (Sampathkumar et al. 2013). CESA complexes are assembled in Golgi apparatus (GA) and are delivered to the PM in vesicles (Crowell et al. 2009; Gutierrez et al. 2009). The delivery of CESA-containing vesicles was impaired and the lifetime of CESA complexes in the PM in double mutants *act2 act7* was longer (Sampathkumar et al. 2013).

Interestingly, there is a bidirectional relationship between cellulose synthesis and actin cytoskeleton; a decrease in AFs dynamics was observed in plants treated with isoxaben, a cellulose synthesis inhibitor (Tolmie et al. 2017). These data clearly suggest that the actin cytoskeleton plays a role in cellulose deposition. However, molecular players linking the actin cytoskeleton to components of the cellulose synthesis machinery are missing. Recently, Zhang et al. (2019) demonstrated the role of myosins XI in the tethering and fusion of vesicles containing CESA complexes to the plasma membrane. Triple myosin mutant *xik xi1 xi2* displays decreased cellulose content, and a detailed analysis shows that AFs depolymerization and myosin inhibition lead to the fusion failure of CESA-containing vesicles with the plasma membrane. Interestingly, inhibition of myosin, but not the actin cytoskeleton, leads to decreased CESA density and motility in the PM, which suggests an actin-independent role of myosin in this process (Zhang et al. 2019). The report of Zhang et al. (2019) thus identified the myosin as the first molecule linking the actin cytoskeleton to CESA function.

1.2.3 Deposition of Noncellulose Polymers

The deposition of noncellulosic polysaccharides is less understood. Pectin and hemicellulose are synthesized in the GA and delivered to the plasma membrane through vesicular transport (Kim and Brandizzi 2014). Vesicles are delivered to the plasma membrane in a cytoskeleton-dependent manner. FRA1 is a kinesin-4 family member, which is believed to transport vesicles originating in the GA to the plasma membrane. FRA1 moves along MTs in a CESA-independent manner, and *fra1* mutants show cell wall defects (Zhu et al. 2015). AFs are also necessary for the delivery of vesicles containing cell wall components to the plasma membrane. Plants with disrupted actin-dependent transport show defects in cell wall deposition and expansion, which has been demonstrated in plants treated with latrunculin B (Baluska et al. 2001), in plants with knocked-out myosins (Peremyslov et al. 2010) and in plants lacking vegetative actin isoforms (Kandasamy et al. 2009; Sampathkumar et al. 2013). However, the molecular mechanism, chiefly an actin-associated protein involved in actin-vesicle interaction, PM targeting, and fusion, remains unknown.

Fine AFs may participate in vesicles fusion with the plasma membrane in a manner similar to that described for tip-growth (Ketelaar 2013). The coordination of cellulose and noncellulosic polymer deposition is also not well understood. Considering the crucial roles of both MTs and AFs in cell wall deposition, it is tempting to speculate that a mechanism linking these two cytoskeletal arrays may be involved in this regulation. Therefore, here we discuss a number of candidate proteins, which are known to cross-link MTs and AFs, and whose function is related to cell wall assembly during diffuse growth.

1.3 Proteins Interacting with Actin and Microtubules with Potential Function in Cell Wall Assembly

1.3.1 Kinesins

Kinesins are molecular motors that convert chemical energy into directional movement along MTs. The family of plant kinesins is exceptionally large, which is considered a consequence of the absence of dynein motors in land plants (Wickstead and Gull 2007). The plant-specific kinesin-14 family contains kinesins with a unique capability to translocate toward the minus end of MTs (Gicking et al. 2018). A subgroup of the kinesin-14 family is comprised of kinesins with a calponin homology (CH) domain, which have been shown to interact with MTs and AFs (KCH subgroup) in species such as cotton (Preuss et al. 2004), rice (Frey et al. 2009; Maruta et al. 2010), *Arabidopsis* (Buschmann et al. 2011), and tobacco (Klotz and Nick 2012). If some of these kinesins aid in cell wall synthesis through MTs and AFs cross-linking, a co-localization with both cytoskeletal systems should be observed in the plant cortex. Indeed, Preuss et al. (2004) demonstrated co-localization of cotton KCH with transverse AFs and Frey et al. (2009) showed OsKCH co-localization with crossover between MTs and AFs in the cell cortex. However, cell wall changes have not been reported for respective mutants. A phenotype associated with altered cell expansion has been reported for mutants lacking tobacco NtKCH (Klotz and Nick 2012), but the NtKCH association with cortical MTs was actin-independent. Therefore, the question of possible participation of these cross-linking kinesins in cell wall assembly remains open.

1.3.2   DREPP/MDP25

DREPP (Developmentally-Regulated Plasma Membrane Polypeptide) proteins represent plant-specific proteins that interact peripherally with the plasma membrane (Gantet et al. 1996; Vosolsobě et al. 2017). The *Arabidopsis* member of the DREPP family, called MDP25, has been shown to bind and destabilize cortical MTs in *Arabidopsis hypocotyls* (Li et al. 2011) and to bind and sever AFs in pollen tubes (Qin et al. 2014) in a Ca^{2+}-dependent manner. Since *mdp25* mutants had longer hypocotyls, and overexpression of MDP25 resulted in shorter hypocotyls, the microtubule-destabilization activity of MDP25 seems to play a role in cell elongation, probably through the control of cortical MTs stability in elongating cells (Li et al. 2011). It has not been demonstrated whether MDP25 function in the cortical region of diffusively expanding cells involves regulation of AFs as well. However, its interaction with cortical MTs, its function in cell elongation, as well as its dual function in the binding of MTs and AFs, make MDP25 another candidate for an actin- and microtubule-interacting protein during cell wall expansion.

1.3.3   Formins

Formins are evolutionary conserved multidomain proteins, which contain a typical FH (formin homology) domain in their structure. One of formin's roles is to promote AFs polymerization. Likewise, some family members also interact with MTs in animal (Bartolini and Gundersen 2010) and plant (Deeks et al. 2010; Wang et al. 2013; Rosero et al. 2013; Sun et al. 2017) cells. Plant formins comprise a large protein family with three clades: Class I, Class II, and Class III (Deeks et al. 2002; Grunt et al. 2008), where Class I often contains transmembrane members, and Class II members peripherally interact with membranes (Cvrčková 2013). Membrane association and the ability to bind MTs and AFs thus make formins a good candidate for the molecular link between both cytoskeletons regarding cell wall assembly control. Moreover, formin mutants often exhibit growth defects. For example, the loss of a Class I AtFH1 protein results in changes in the MT and AF cytoskeletons and altered cell expansion (Rosero et al. 2013, 2016). Rice Class II formin FH5 binds AFs and MTs, and functional analysis of respective mutants proved the role of FH5 in cell expansion (Zhang et al. 2011a). Sun et al. (2017) demonstrated that rice formin OsFH5 is involved in actin assembly, and the binding and cross-linking of AFs and MTs. The overexpression of OsFH5 results in the formation of larger grains due to altered cell expansion. Here the dense actin network formed in overexpressing plants probably promoted cell expansion, presumably through stabilization of the plasma membrane-associated cytoskeleton and speeded cell wall assembly (Sun et al. 2017). The role in cross-linking of AFs and MTs has been demonstrated also for formin AtFH4 (Deeks et al. 2010) and Class II formin AthFH16 (Wang et al. 2013). Since no cell wall defects have been reported for plants lacking these formins, it is not known whether they participate in cell wall assembly. During cell division, a specific role is played by the class II formin AFH14. AFH14 was shown to interact with both AFs and MTs. The T-DNA insertion mutant *afh14* of *Arabidopsis* shows defects in meiosis and tetrad formation, suggesting a specific role during cell division (Li et al. 2010). In summary, at least AtFH1, rice FH5, and OsFH15 formins clearly link the control of AFs and MTs regulation with cell wall assembly. Although the mechanism needs to be elucidated, the control of cortical cytoskeleton seems to be tightly connected with formin function. Importantly, AtFH1 was shown to connect the cell wall with the plasma membrane and the cortical cytoskeleton, and its localization in the plasma membrane to specific domains was restricted by MTs (Martinière et al. 2011), which may be the mechanism of communication within the cytoskeleton–plasma membrane–cell wall continuum.

1.3.4 Arp2/3

Arp2/3 is an evolutionarily conserved protein complex, whose main function is the nucleation of AFs (Welch et al. 1998). The complex consists of 7 subunits (Welch et al. 1997), which are also encoded in plant genomes. *Arabidopsis* plants lacking a functional Arp2/3 complex display a broad spectrum of phenotypes, which suggests a specific role of Arp2/3-controlled actin nucleation in cell wall building. Defects in cell–cell adhesion have been reported for cotyledons and etiolated hypocotyls of mutants lacking Arp2/3 complex subunits (Le et al. 2003; Mathur et al. 2003a, b; El-Assal et al. 2004; Kotchoni et al. 2009; Zhang et al. 2013; Pratap Sahi et al. 2017). The hallmark phenotype of Arp2/3 mutants is distorted trichomes (Szymanski et al. 1999; Mathur et al. 1999; Schwab et al. 2003; Le et al. 2003; Li et al. 2003; El-Assal et al. 2004; Basu et al. 2004, 2005), which fail to form narrow and pointed trichome branches in the absence of an active Arp2/3 complex. As demonstrated by

Yanagisawa et al. (2015), the distorted phenotype is based on aberrant cell wall deposition within the growing apex. In the work of Sahi et al. (2017), mutants lacking several Arp2/3 subunits were shown to deposit cell walls of altered composition in inflorescence stems; these cell walls contained less cellulose and more pectin. Interestingly, several reports noted microtubular changes in Arp2/3 mutants, which indicated that the Arp2/3 complex may also be involved in the regulation of MTs (Saedler et al. 2004a; Zhang et al. 2005). A direct interaction of Arp2/3 with MTs was suggested by Havelková et al. (2015), who showed that the Arp2/3 complex subunit ARPC2 binds to MTs. Since the Arp2/3-nucleated actin cytoskeleton is clearly involved in the control of cell wall deposition, despite the mechanism not being clear, its interaction with MTs makes the Arp2/3 complex another candidate in MT-AF cross-talk during cell wall assembly.

1.4 Patterning for Polar Cell Wall Deposition in Diffusively Growing Cells: Cooperation by Restriction?

Cellulose synthesis and deposition of noncellulosic polymers must be coordinated in isotropically expanding plant cells. Our knowledge of cellulose synthesis and its cytoskeletal control has increased considerably in recent years due to several important findings, but the mechanism and control of noncellulosic polymer deposition during cell wall expansion is less understood. In many cases, diffusively growing plant cells deposit the cell wall in a polarized manner, which allows them to achieve complex shapes. Several cell types that undergo complex morphogenesis became models for research in cytoskeletal control of cell deposition.

Current data suggest that during polarized cell wall deposition and expansion, specific mechanisms are often activated based on the cooperative restriction of both cytoskeletons, rather than direct cross-linking. Several examples indicate that AFs and MTs may delineate cortical regions to enable polarized cell expansion by differential deposition of cell wall. This leads to the formation of cell wall domains, with distinct mechanical properties that respond differentially to turgor pressure, thus enabling various cell shapes. This concept is supported by recently uncovered molecular mechanisms that control the differential deposition of the cell wall in plant cells with complex shapes, as reviewed later in this chapter. Perhaps understanding the specific roles of AFs and MTs in cell wall deposition in these specialized cases will help to uncover the general mechanism coordinating cellulose and noncellulose polysaccharide deposition in diffusively expanding cells.

1.4.1 Developing Trichomes and Cytoskeletal Patterning of Cell Wall Deposition

The unicellular trichomes of Arabidopsis leaves are a classical model of plant morphogenesis (Hulskamp et al. 1998). Trichome development is under the control of both the microtubular and actin cytoskeleton (Szymanski 2009). MTs contribute to trichome branch initiation and trichome growth during anisotropic cell expansion of the main stalk (Mathur and Chua 2000; Folkers et al. 2002). To direct cell expansion, cortical MTs are oriented perpendicularly to the direction of growth, forming a collar (Sambade et al. 2014). Close inspection suggests that MTs are involved in branch marking, but they are not needed for bulge formation (Sambade et al. 2014). Actin has been shown present at the tip of newly emerging trichomes and trichome branches. Data indicate that, in the context of trichome branching, actin would be crucial to determining future sites of MTs localization by affecting microtubule dynamics (Schwab et al. 2003; Saedler et al. 2004a; Sambade et al. 2014). During branch elongation, the Arp2/3 complex plays a specific role, as distorted trichomes are typical phenotypes of mutants lacking functional Arp2/3 (see abovementioned roles of Arp2/3 and references there). The role of the Arp2/3 complex in trichome branch growth was analyzed in detail by Yanagisawa et al. (2015). The authors demonstrated that during trichome branch elongation, cell wall of gradually increasing thickness is deposited at the growing branch tip, which endows the trichome cell wall with specific biomechanical properties. The Arp2/3-nucleated actin cytoskeleton at the branch tip aids in polar cell wall deposition, because mutants defective in activated Arp2/3 failed to form pointed trichome structure (Yanagisawa et al. 2015). A detailed mechanism of the Arp2/3-nucleated AFs in branch elongation was shown by Yanagisawa et al. (2018). SPIKE1 is a protein that functions as a GEF, which also controls Arp2/3 complex activation through the Arp2/3-activating complex WAVE/SCAR (Basu et al. 2008). SPIKE1 localizes to the trichome branch tip, where it is concentrated and stabilized during the course of trichome development. SPIKE1 restriction to the branch tip is controlled by microtubular bundles found on the flank, but absent in the tips of growing trichomes (Yanagisawa et al. 2018). MTs in the growing trichome branches thus develop a microtubule-depleted zone (MDZ) in the tip (Yanagisawa et al. 2015). SPIKE1 patches are destabilized in the vicinity of MTs and have a shorter lifetime than SPIKE1 in MDZ, suggesting its stabilization in the tip of the branch. The stabilized population of SPIKE1 then marks the place of WAVE/SCAR complex activation, Arp2/3 activation, and AFs assembly (Yanagisawa et al. 2018). The authors presumed that the polymerized actin mediates the flow of cytoplasm and the delivery of vesicles containing cell wall material into the tip of growing trichome branches. Here, a specialized cell wall domain with a characteristic thickness gradient is believed to enable proper trichome development. The developing trichome branch thus represents an example of MTs restricting AFs polymerization for differential cell wall deposition, as is necessary for polarized growth.

1.4.2   Leaf Epidermal Cells

Leaf epidermal cells are an outstanding example of plant cell shape control. Through a multistep morphogenic process, they develop from polygon-shaped cells into large cells with a complex interlocking pattern of lobes and necks, similar to that of a puzzle (Vőfély et al. 2019).

Careful time-lapse investigations following cell morphogenesis indicate that pavement cells go through a phase of lobe initiation followed by a phase of cell expansion (Armour et al. 2015; Zhang et al. 2011b). This set of subcellular responses is thought to be regulated by tissue-wide molecular and mechanical stimuli. Both MTs and AFs are involved in the control of pavement cell shape, and the current hypothesis is based on their mutual interaction during cell expansion and cell wall building. Preliminary observations hinted at microtubule organization and cellulosic cell wall deposition as main players in PC morphogenesis. Briefly, cellulose microfibrils were visualized at the periclinal–anticlinal wall junctions of neck regions, which continued all the way down the anticlinal wall (Panteris et al. 1993a, b, 1994) mirroring microtubule orientation (Panteris et al. 1993a, b, 1994; Kirik et al. 2007; Fu et al. 2002, 2005; Zhang et al. 2011b). To date, visualization of MTs along anticlinal and periclinal cell surfaces during the initiation of lobe formation has yielded contradictory conclusions. While Armour et al. (2015) observed that a higher occupancy of MTs in anticlinal and periclinal walls was a predictor of lobe formation, Belteton et al. (2018) could not correlate the presence of MTs with the emergence or existence of lobes. Since MTs are highly dynamic structures, higher spatiotemporal resolution will be of utmost importance to elucidate their role during cell morphogenesis. AFs were introduced into the model, indicating the need for them in lobe elongation. Mutants lacking a functional Arp2/3 complex (Basu et al. 2004, 2005, 2008; Brembu et al. 2004; Frank et al. 2003; Frank and Smith 2002; Le et al. 2003, 2006; Li et al. 2003; Mathur et al. 2003a, b; Qiu et al. 2002; Saedler et al. 2004b) were shown to develop reduced lobes in pavement cells. Actin has been detected in lobes (Fu et al. 2002, 2005; Armour et al. 2015), and the involvement of the cytoskeleton in the process was further confirmed in pharmacological studies, where depolymerization of MTs (Akita et al. 2015) and AFs (Rosero et al. 2016) impacts pavement cell morphogenesis. A molecular model was established involving Rho GTPase of plants (ROP) signaling cascade controlling lobes formation through the control of cortical cytoskeleton. In this model, activated ROP2/ROP4 triggers the activity of ROP-interactive CRIB motif-containing protein 4 (RIC4), promoting the formation of dynamic AFs in the lobe outgrowth (Fu et al. 2005). In this context, AFs were suggested to participate in the delivery of vesicles containing new cell wall material or cell wall loosening enzymes to the elongating lobes (Smith 2003). Furthermore, activation of ROP6 localized at the neck regions would induce the rearrangement of MTs at the sites of growth restriction via the action of its RIC1 effector (Fu et al. 2009), which activates KATANIN protein with microtubule-severing activity (Lin et al. 2013). Spatial restriction of actin and microtubular cytoskeletons at the cell cortex would be mediated through the ROP2 molecule, which activates RIC4 in the lobe, but inhibits RIC1–KATANIN mediated microtubule alignments in the neck regions (Fu et al. 2005; Lin et al. 2013; Li et al. 2017).

Initial work proposed a model in which auxin played a central role. In the model, auxin activated the ROP2 pathway via the ABP1–TMK auxin receptor complex. This signaling cascade promoted PIN1 accumulation in lobe tips, which induced auxin accumulation in the extracellular space, generating a positive feedback loop (Xu et al. 2010, 2011, 2014; Chen and Yang 2014; Chen et al. 2015; Guo et al. 2015). However, this view has been put into question due to recent evidence showing newly generated *abp1* lines lacking a pavement cell phenotype (Gao et al. 2015) and the analysis of in vivo localization of PIN auxin carriers in combination with the lack of a lobe phenotype in higher order *PIN* null allele mutants (Belteton et al. 2018). Nonetheless, auxin could have a role in pavement cell expansion and extend the period of lobe formation, as is suggested by long-term treatments with low levels of auxin (Xu et al. 2010; Gao et al. 2015; Belteton et al. 2018).

The alternating pattern of lobes and indentations may arise from locally modified mechanical properties of growing cell walls, which respond differentially to stress forces during growth. Indeed, cellulose microfibril deposition in periclinal–anticlinal wall junctions of necks would strengthen the cell wall against expansion in the model (Sapala et al. 2018). Other analyses have evidenced the need for differential cell wall composition and active remodeling in order to produce interdigitated cells (Majda et al. 2017; Sotiriou et al. 2018). Particularly, low methylesterification of homogalacturonan (HG) was proven to be a predictor of future lobe formation very early in plant development (Majda et al. 2017). The mechanism of cytoskeletal contribution to cell wall remodeling remains to be demonstrated. The role of microtubular rearrangement seems to lie in the deposition of cellulose in regions where the cell wall must resist mechanical stress (e.g. turgor pressure). The role of actin is far from understood. Perhaps actin localized to lobes is responsible for the delivery of vesicles with wall material and cell wall modifying enzymes. However, Armour et al. (2015) noted that no specific accumulation of actin occurs during lobe initiation, yet pharmacological studies with AFs depolymerizing drugs suggest a role of actin in this process (Armour et al. 2015). Arp2/3 mutants display pavement cell lobe formation defects. Importantly, the ROP-GEF protein SPIKE1 interacts with NAP1, a SCAR/WAVE complex activating Arp2/3 (Basu et al. 2008), which directly connects the ROP signaling pathway with actin remodeling. However, the role of SPIKE1-SCAR/WAVE-Arp2/3 in the signal pathway of ROP2-RIC4 needs further investigation.

Recent studies indicated the importance of the mechanics of the cell wall itself. A feedback loop exists between the cytoskeleton and the cell wall, where stimuli from the cell wall also control the cytoskeleton. Sampathkumar et al. (2014) demonstrated that MTs respond to cell wall stress patterns. Adhesion-defective mutants support the existence of the feedback loop, where cell adhesion properties, maintained by heterogeneous pectin esterification, are required to spread stress forces crucial to sustain tension-induced cell wall reinforcement (Verger et al. 2018). Therefore, the achievement of the interlocked pattern displayed by pavement cells is a result of the coordination in time and space between molecular interactions, cytoskeletal organization, and cell wall deposition.

1.4.3   Xylem Development: An Emerging New Site for Microtubule–Actin–AF Coordination in Cell Wall Deposition

Xylem vessel cells are specialized cells that develop their secondary cell wall in distinct patterns, such as spiral, annular, or reticular. After secondary cell wall deposition, xylem vessel cells undergo programmed cell death, leaving only the thickened cell walls, which form conductive tubes for water transport through the plant body. Early observations identified cortical microtubular bundles underlying cell wall deposition regions (Hepler and Newcomb 1964), and the central role of MTs in the pattern localization of secondary cell wall is now well established (Roberts et al. 2004; Oda 2005). The mechanism underlying patterned secondary cell wall formation is based on the local formation of microtubular bundles under the plasma membrane, which direct CESA complexes (Paredez et al. 2006). The formation of xylem vessel cell wall patterning requires local destabilization and polymerization of cortical MTs on the plasma membrane, which is based on a positive feedback loop of the signaling cascade involving ROP GTPase. During xylem vessel development, ROPGEF4 anchored to the plasma membrane locally and spontaneously activates ROP11, the central regulator of localized cell wall deposition. The GTPase activating factor ROPGAP3 in turn inactivates ROP11. As a result, ROP11 is locally activated at the plasma membrane, and this self-organization mechanism allows for the formation of evenly spaced plasma membrane domains with active ROP11 (Oda and Fukuda 2013b). ROP11 localized in these domains recruits MICROTUBULE-DEPLETION DOMAIN 1 (MIDD1) protein to the plasma membrane. MIDD1 is preferentially expressed in developing xylem cells and it binds MTs and the plasma membrane through its two coiled-coil domains (Oda and Fukuda 2012). MIDD1 recruits the AtKinesin-13A protein to the cortical MTs (Oda and Fukuda 2013a). AtKinesin-13A demonstrates microtubule depolymerizing activity, and its recruitment to the plasma membrane domains of activated ROP11 locally destabilizes MTs. Therefore, ROP11 plasma membrane domains are devoid of MTs, while MTs out of these domains are stabilized by other proteins, such as MAP70, which localizes along borders of microtubular bundles between formed pits (Pesquet et al. 2010). A number of other proteins were demonstrated to participate in establishing plasma membrane domains for secondary cell wall deposition. For example, an IQD13 protein, member of the IQ67 DOMAIN (IQD) protein family (Abel et al. 2013) which contains many microtubule-binding members (Bürstenbinder et al. 2017), binds MTs and the plasma membrane to limit lateral diffusion of ROP11 domains (Sugiyama et al. 2017). Recently, the actin cytoskeleton was added to the list of proteins participating in xylem vessel cell morphogenesis. At pit boundaries, bordered cell walls develop. Their formation has been shown to be controlled by F-actin rings assembled around pits (Sugiyama et al. 2019). Interestingly, actin assembly at pit boundaries is controlled by a distinct signaling pathway involving ROP11. Actin polymerization is promoted by WALLIN (WAL), which is recruited to pit borders by Boundary of ROP Domain1 (BDR1) protein. BDR1 is a ROP effector and is recruited to pit boundaries by active ROP (Sugiyama et al. 2019). This recently described pathway suggests that xylem vessel cell formation is another example of the coordinated regulation of actin and microtubular cytoskeletons in the cortical domain to assist in different cell wall domain synthesis.

1.4.4   Preprophase Band: A Place for Direct Cooperation?

Here we discuss also a specialized cytoskeletal structure of great morphogenetic importance, which is assembled in the plant cortex. The preprophase band (PPB), which is made up of MTs and AFs, is not involved in the deposition of the cell wall. However, this cytoskeletal structure positions molecular markers on the plasma membrane, which are later recognized during cytokinesis as the site for fusion of maternal and newly synthesized cell wall. The PPB is present in most land plants and its origin is probably related to the necessity of precise control regarding cell division polarity (Buschmann and Zachgo 2016). Current reports suggest that the PPB functions as a noise-reducing mechanism contributing to the proper spatial control of plant cell divisions (Schaefer et al. 2017). Since the formation of the microtubular PPB has been shown to be aided by cortical AFs, we are mentioning PPB formation as a specific example of microtubule–actin filament interaction in the cortical cytoplasm.

The PPB is a microtubular structure, formed through subsequent reorganization of cortical MTs into a microtubular ring which encircles the cell in the cortical zone prior to mitosis. As a narrow ring, the PPB delimits a domain in the cortical cytoplasm called the cortical division zone (CDZ) (Rasmussen et al. 2013). A number of microtubule-associated proteins participate in PPB formation (McMichael and Bednarek 2013). The TTP protein complex coordinates microtubular reorganization in the PPB. The TTP complex consists of TONNEAU1 (TON1) protein, which is similar to human centrosomal proteins, PP2 phosphatase with regulatory subunits FASS/TON2 (Spinner et al. 2013), and the protein TRM1 (TON-recruiting motif1) which recruits TON1 to the MTs and belongs to a family of TRM proteins (Drevensek et al. 2012). Another group of proteins is positioned on the plasma membrane in a microtubule-dependent manner, thus marking the site for future fusion of newly synthesized cell wall. These include the kinesin-12 proteins PHRAGMOPLAST-ORIENTING KINESIN (POK) 1 and POK2 (Lipka et al. 2014), TANGLED1 (Walker et al. 2007), RanGAP1 (Xu et al. 2008), AUXIN-INDUCED IN ROOT CULTURES (AIR9) (Buschmann et al. 2006), and a minus-end directed calmodulin-binding kinesin-14 KCBP/ZWICHEL (ZWI) (Buschmann et al. 2015). These molecules are components of the "memory mechanism" that marks the CDZ. Interactions and specific roles of CDZ markers during the cell division are partially understood and are described elsewhere (Müller and Jürgens 2016). At the time of nuclear envelope breakdown, PPB MTs depolymerize and disappear completely from the cortical zone. However, the identity of CDZ is maintained thanks to marker molecules, which mostly remain associated with the PM during mitosis. Later during telophase, the CDZ is recognized by a plant cytokinetic apparatus called the phragmoplast as the site of fusion of newly built, centrifugally growing cell plate with the maternal cell wall. Thus, the formation of the microtubular PPB in the cortical zone is a mechanism through which the polarity of cell division is established before the division itself (Müller and Jürgens 2016).

Actin is localized in the PPB (Traas et al. 1987; Palevitz 1987; Mineyuki and Palevitz 1990; Cleary et al. 1992). However, the role of AFs in PPB formation is less understood. Actin co-localizes with MTs in the PPB as a broad band that narrows together with the MTs. When the PPB reaches its most narrow configuration, actin starts to disappear from the cortical zone, forming an actin-depleted zone (ADZ) (Cleary et al. 1992; Liu and Palevitz 1992; Cleary 1995). In BY-2 cells, the ADZ is bordered by an actin-rich zone, the so-called "twin peaks" (Sano et al. 2005). While PPB MTs depolymerize at the onset of mitosis, the ADZ remains a negative marker of the CDZ in the cortical region throughout mitosis. AFs and MTs have been demonstrated to interact in the PPB because the depolymerization of MTs results in actin localization changes (Palevitz 1987). Actin assists in the process of PPB narrowing, which is impaired if actin is disrupted (Mineyuki and Palevitz 1990; Eleftheriou and Palevitz 1992). Depolymerization of the AFs results in misoriented cell division (Mineyuki and Palevitz 1990; Eleftheriou and Palevitz 1992; Granger and Cyr 2001; Hoshino et al. 2003), which suggests that actin has a role in the positioning of new cell wall either during PPB formation or later during cell wall fusion. In vivo analysis of the ADZ revealed that the ADZ actually contains scarce AFs rather than being completely devoid of actin (Panteris 2008). The role of actin–microtubule interactions during PPB narrowing was strongly confirmed by the observations of Takeuchi et al. (2016). Short, single AFs were observed in the PPB using electron tomography, most of which were bound to MTs. During later stages of microtubular PPB formation, actin–microtubular bridges were displaced by microtubule–microtubule interactions (Takeuchi et al. 2016). Therefore, short AFs may cross-link MTs during PPB formation and bring them together in order to enable the formation of microtubular bundles. As the narrow structure of the microtubular PPB occupies the cell cortex, the actin cytoskeleton is removed to form the ADZ. Two types of microtubular cross-links were identified using electron tomography: longer cross-linkers that dominated during the early stages of PPB formation, and shorter cross-linkers that were mostly observed during PPB narrowing (Takeuchi et al. 2016). The identities of the proteins connecting AFs and MTs in the PPB are yet to be identified. Several candidate proteins exist, which may mediate aforementioned interactions in the PPB. Members of the kinesin-14 family, which interact with both MTs and AFs, have been reported to localize to the PPB (Klotz and Nick 2012; Buschmann et al. 2015). Recently, the minus-end-directed kinesin 14 OsKCH2 has been shown to decorate PPB in rice cells. OsKCH2 also binds AFs and transports AFs along MTs in vitro (Tseng et al. 2018). Furthermore, type II formin 14 AFH14, which binds AFs and MTs, has been shown to localize to the PPB (Li et al. 2010). Localization of other actin–microtubule interacting proteins in the PPB remains to be tested.

1.5   Conclusion

MTs and AFs cross-talk is mediated by a number of molecular interactors in plant cells. After close inspection, the functions of only a few of these proteins seem to be related to cell wall building. Lessons from distinct microtubule and actin functions in localized cell wall deposition in cells with complex shapes suggest that it is interconnected regulatory pathways, rather than a direct cross-link, that mediate cytoskeletal interactions. Here, MTs and AFs often cooperate through mutual restriction, controlled by signaling pathways with shared components. MTs and AFs delimit differential zones on the plasma membrane for the formation of distinct cell wall domains. The building of the cytoskeletal structure of the preprophase band, a specialized cytoskeletal structure, represents an example of direct MTs and AFs interaction in the cortical cytoplasm.

Acknowledgments Authors would like to thank Lena Hunt for manuscript proofreading. Financial support: NPU I, LO1417 (Ministry of Education, Youth and Sports of the Czech Republic).

References

Abel S, Bürstenbinder K, Müller J (2013) The emerging function of IQD proteins as scaffolds in cellular signaling and trafficking. Plant Signal Behav 8:e24369

Akita K, Higaki T, Kutsuna N, Hasezawa S (2015) Quantitative analysis of microtubule orientation in interdigitated leaf pavement cells. Plant Signal Behav 10:e1024396

Armour WJ, Barton DA, Law AMK, Overall RL (2015) Differential growth in periclinal and anticlinal walls during lobe formation in Arabidopsis cotyledon pavement cells. Plant Cell 27:2484–2500

Baluska F, Jasik J, Edelmann HG, Salajova T, Volkmann D (2001) Latrunculin B-induced plant dwarfism: plant cell elongation is F-actin-dependent. Dev Biol 231:113–124

Bartolini F, Gundersen GG (2010) Formins and microtubules. Biochim Biophys Acta Mol Cell Res 1803:164–173

Bashline L, Lei L, Li S, Gu Y (2014) Cell wall, cytoskeleton, and cell expansion in higher plants. Mol Plant 7:586–600

Basu D, El-Assal SE-D, Le J, Mallery EL, Szymanski DB (2004) Interchangeable functions of Arabidopsis PIROGI and the human WAVE complex subunit SRA1 during leaf epidermal development. Development 131:4345–4355

Basu D et al (2005) DISTORTED3/SCAR2 is a putative arabidopsis WAVE complex subunit that activates the Arp2/3 complex and is required for epidermal morphogenesis. Plant Cell 17:502–524

Basu D, Le J, Zakharova T, Mallery EL, Szymanski DB (2008) A SPIKE1 signaling complex controls actin-dependent cell morphogenesis through the heteromeric WAVE and ARP2/3 complexes. Proc Natl Acad Sci U S A 105:4044–4049

Belteton SA, Sawchuk MG, Donohoe BS, Scarpella E, Szymanski DB (2018) Reassessing the roles of PIN proteins and anticlinal microtubules during pavement cell morphogenesis. Plant Physiol 176:432–449

Brembu T, Winge P, Seem M, Bones AM (2004) NAPP and PIRP encode subunits of a putative wave regulatory protein complex involved in plant cell morphogenesis. Plant Cell 16:2335–2349

Bringmann M, Li E, Sampathkumar A, Kocabek T, Hauser M-T, Persson S (2012) POM-POM2/ CELLULOSE SYNTHASE INTERACTING1 is essential for the functional association of cellulose synthase and microtubules in Arabidopsis. Plant Cell 24:163–177

Bürstenbinder K, Möller B, Plötner R, Stamm G, Hause G, Mitra D, Abel S (2017) The IQD family of calmodulin-binding proteins links calcium signaling to microtubules, membrane subdomains, and the nucleus. Plant Physiol 173:1692–1708

Buschmann H, Zachgo S (2016) The evolution of cell division: from Streptophyte algae to land plants. Trends Plant Sci 21:872–883

Buschmann H, Chan J, Sanchez-Pulido L, Andrade-Navarro MA, Doonan JH, Lloyd CW (2006) Microtubule-associated AIR9 recognizes the cortical division site at preprophase and cell-plate insertion. Curr Biol 16:1938–1943

Buschmann H, Green P, Sambade A, Doonan JH, Lloyd CW (2011) Cytoskeletal dynamics in interphase, mitosis and cytokinesis analysed through agrobacterium-mediated transient transformation of tobacco BY-2 cells. New Phytol 190:258–267

Buschmann H, Dols J, Kopischke S, Pena EJ, Andrade-Navarro MA, Heinlein M, Szymanski DB, Zachgo S, Doonan JH, Lloyd CW (2015) Arabidopsis KCBP interacts with AIR9 but stays in the cortical division zone throughout mitosis via its MyTH4-FERM domain. J Cell Sci 128:2033–2046

Chen J, Yang Z (2014) Novel ABP1-TMK auxin sensing system controls ROP GTPase-mediated interdigitated cell expansion in Arabidopsis. Small GTPases 5:e29711

Chen J, Wang F, Zheng S, Xu T, Yang Z (2015) Pavement cells: a model system for non-transcriptional auxin signalling and crosstalks: Fig. 1. J Exp Bot 66:4957–4970

Cleary AL (1995) F-actin redistributions at the division site in living Tradescantia stomatal complexes as revealed by microinjection of rhodamine-phalloidin. Protoplasma 185:152–165

Cleary AL, Gunning BES, Wasteneys GO, Hepler PK (1992) Microtubule and F-actin dynamics at the division site in living Tradescantia stamen hair cells. J Cell Sci 103:977–988

Cosgrove DJ (2005) Growth of the plant cell wall. Nat Rev Mol Cell Biol 6:850–861

Crowell EF, Bischoff V, Desprez T, Rolland A, Stierhof Y-D, Schumacher K, Gonneau M, Hofte H, Vernhettes S (2009) Pausing of Golgi bodies on microtubules regulates secretion of cellulose synthase complexes in Arabidopsis. Plant Cell 21:1141–1154

Cvrčková F (2013) Formins and membranes: anchoring cortical actin to the cell wall and beyond. Front Plant Sci 4:436

Deeks MJ, Hussey PJ, Davies B (2002) Formins: intermediates in signal-transduction cascades that affect cytoskeletal reorganization. Trends Plant Sci 7:492–498

Deeks MJ, Fendrych M, Smertenko A, Bell KS, Oparka K, Cvrckova F, Zarsky V, Hussey PJ (2010) The plant formin AtFH4 interacts with both actin and microtubules, and contains a newly identified microtubule-binding domain. J Cell Sci 123:1209–1215

Drevensek S, Goussot M, Duroc Y, Christodoulidou A, Steyaert S, Schaefer E, Duvernois E, Grandjean O, Vantard M, Bouchez D, Pastuglia M (2012) The Arabidopsis TRM1-TON1 interaction reveals a recruitment network common to plant cortical microtubule arrays and eukaryotic centrosomes. Plant Cell 24:178–191

El-Assal SED, Le J, Basu D, Mallery EL, Szymanski DB (2004) Distorted2 encodes an ARPC2 subunit of the putative Arabidopsis ARP2/3 complex. Plant J 38:526–538

Eleftheriou EP, Palevitz BA (1992) The effect of cytochalasin D on preprophase band organization in root tip cells of Allium. J Cell Sci 103:989–998

Endler A, Schneider R, Kesten C, Lampugnani ER, Persson S (2016) The cellulose synthase companion proteins act non-redundantly with CELLULOSE SYNTHASE INTERACTING1/ POM2 and CELLULOSE SYNTHASE 6. Plant Signal Behav 11:1–4

Folkers U, Kirik V, Schöbinger U, Falk S, Krishnakumar S, Pollock MA, Oppenheimer DG, Day I, Reddy AR, Jürgens G, Hülskamp M (2002) The cell morphogenesis gene ANGUSTIFOLIA encodes a CtBP/BARS-like protein and is involved in the control of the microtubule cytoskeleton. EMBO J 21:1280–1288

Frank MJ, Smith LG (2002) A small, novel protein highly conserved in plants and animals promotes the polarized growth and division of maize leaf epidermal cells. Curr Biol 12:849–853

Frank MJ, Cartwright HN, Smith LG (2003) Three brick genes have distinct functions in a common pathway promoting polarized cell division and cell morphogenesis in the maize leaf epidermis. Development 130:753–762

Frey N, Klotz J, Nick P (2009) Dynamic bridges—a calponin-domain kinesin from rice links actin filaments and microtubules in both cycling and non-cycling cells. Plant Cell Physiol 50:1493–1506

Fu Y, Li H, Yang ZB (2002) The ROP2 GTPase controls the formation of cortical fine F-actin and the early phase of directional cell expansion during Arabidopsis organogenesis. Plant Cell 14:777–794

Fu Y, Gu Y, Zheng ZL, Wasteneys G, Yang ZB (2005) Arabidopsis interdigitating cell growth requires two antagonistic pathways with opposing action on cell morphogenesis. Cell 120:687–700

Fu Y, Xu T, Zhu L, Wen M, Yang Z (2009) A ROP GTPase signaling pathway controls cortical microtubule ordering and cell expansion in Arabidopsis. Curr Biol 19:1827–1832

Gantet P, Masson F, Domergue O, Marquis-Mention M, Bauw G, Inze D, Rossignol M, de la Serve BT (1996) Cloning of a cDNA encoding a developmentally regulated 22 kDa polypeptide from tobacco leaf plasma membrane. Biochem Mol Biol Int 40:469–477

Gao Y, Zhang Y, Zhang D, Dai X, Estelle M, Zhao Y (2015) Auxin binding protein 1 (ABP1) is not required for either auxin signaling or Arabidopsis development. Proc Natl Acad Sci 112:2275–2280

Gicking AM, Swentowsky KW, Dawe RK, Qiu W (2018) Functional diversification of the kinesin-14 family in land plants. FEBS Lett 592:1918–1928

Granger C, Cyr R (2001) Use of abnormal preprophase bands to decipher division plane determination. J Cell Sci 114:599–607

Grunt M, Žárský V, Cvrčková F (2008) Roots of angiosperm formins: the evolutionary history of plant FH2 domain-containing proteins. BMC Evol Biol 8:1–19

Guo X, Qin Q, Yan J, Niu Y, Huang B, Guan L, Li Y, Ren D, Li J, Hou S (2015) TYPE-ONE PROTEIN PHOSPHATASE4 regulates pavement cell interdigitation by modulating PIN-FORMED1 polarity and trafficking in Arabidopsis. Plant Physiol 167:1058–1075

Gutierrez R, Lindeboom JJ, Paredez AR, Emons AMC, Ehrhardt DW (2009) Arabidopsis cortical microtubules position cellulose synthase delivery to the plasma membrane and interact with cellulose synthase trafficking compartments. Nat Cell Biol 11:797–806

Havelková L, Nanda G, Martinek J, Bellinvia E, Sikorová L, Šlajcherová K, Seifertová D, Fischer L, Fišerová J, Petrášek J, Schwarzerová K (2015) Arp2/3 complex subunit ARPC2 binds to microtubules. Plant Sci 241:96–108

Hepler PK, Newcomb EH (1964) Microtubules and fibrils in the cytoplasm of Coleus cells undergoing secondary wall deposition. J Cell Biol 20:529–532

Hoshino H, Yoneda A, Kumagai F, Hasezawa S (2003) Roles of actin-depleted zone and preprophase band in determining the division site of higher-plant cells, a tobacco BY-2 cell line expressing GFP-tubulin. Protoplasma 222:157–165

Hulskamp M, Schnittger A, Folkers U (1998) Pattern formation and cell differentiation: trichomes in Arabidopsis as a genetic model system. Intl Rev Cytol 186:147–178

Kandasamy MK, McKinney EC, Meagher RB (2009) A single vegetative actin isovariant overexpressed under the control of multiple regulatory sequences is sufficient for normal Arabidopsis development. Plant Cell 21:701–718

Kesten C et al (2019) The companion of cellulose synthase 1 confers salt tolerance through a tau-like mechanism in plants. Nat Commun 10:857

Ketelaar T (2013) The actin cytoskeleton in root hairs: all is fine at the tip. Curr Opin Plant Biol 16:749–756

Kim SJ, Brandizzi F (2014) The plant secretory pathway: an essential factory for building the plant cell wall. Plant Cell Physiol 55:687–693

Kirik V, Herrmann U, Parupalli C, Sedbrook JC, Ehrhardt DW, Hülskamp M (2007) CLASP localizes in two discrete patterns on cortical microtubules and is required for cell morphogenesis and cell division in Arabidopsis. J Cell Sci 120:4416–4425

Klotz J, Nick P (2012) A novel actin-microtubule cross-linking kinesin, NtKCH, functions in cell expansion and division. New Phytol 193:576–589

Kotchoni SO, Zakharova T, Mallery EL, Le J, El-Assal SE-D, Szymanski DB (2009) The association of the Arabidopsis actin-related protein2/3 complex with cell membranes is linked to its assembly status but not its activation. Plant Physiol 151:2095–2109

Krtková J, Benáková M, Schwarzerová K (2016) Multifunctional microtubule-associated proteins in plants. Front Plant Sci 7:474

Le J, El-Assal SE-D, Basu D, Saad ME, Szymanski DB, El Assal SED, Basu D, Saad ME, Szymanski DB (2003) Requirements for Arabidopsis ATARP2 and ATARP3 during epidermal development. Curr Biol 13:1341–1347

Le J, Mallery EL, Zhang C, Brankle S, Szymanski DB (2006) Arabidopsis BRICK1/HSPC300 is an essential WAVE-complex subunit that selectively stabilizes the Arp2/3 activator SCAR2. Curr Biol 16:895–901

Li S, Blanchoin L, Yang Z, Lord EM (2003) The putative Arabidopsis arp2/3 complex controls leaf cell morphogenesis. Plant Physiol 132:2034–2044

Li Y, Shen Y, Cai C, Zhong C, Zhu L, Yuan M, Ren H (2010) The type II Arabidopsis formin14 interacts with microtubules and microfilaments to regulate cell division. Plant Cell 22:2710–2726

Li J, Wang X, Qin T, Zhang Y, Liu X, Sun J, Zhou Y, Zhu L, Zhang Z, Yuan M, Mao T (2011) MDP25, a novel calcium regulatory protein, mediates hypocotyl cell elongation by destabilizing cortical microtubules in Arabidopsis. Plant Cell 23:4411–4427

Li S, Lei L, Somerville CR, Gu Y (2012) Cellulose synthase interactive protein 1 (CSI1) links microtubules and cellulose synthase complexes. Proc Natl Acad Sci 109:185–190

Li C, Lu H, Li W, Yuan M, Fu Y (2017) A ROP2-RIC1 pathway fine-tunes microtubule reorganization for salt tolerance in Arabidopsis. Plant Cell Environ 40:1127–1142

Lin D, Cao L, Zhou Z, Zhu L, Ehrhardt D, Yang Z, Fu Y (2013) Rho GTPase signaling activates microtubule severing to promote microtubule ordering in Arabidopsis. Curr Biol 23:290–297

Lipka E, Gadeyne A, Stöckle D, Zimmermann S, De Jaeger G, Ehrhardt DW, Kirik V, Van Damme D, Müller S (2014) The Phragmoplast-Orienting Kinesin-12 class proteins translate the positional information of the preprophase band to establish the cortical division zone in Arabidopsis thaliana. Plant Cell 26:2617–2632

Liu B, Palevitz BA (1992) Organization of cortical microfilaments in dividing root cells. Cell Motil Cytoskeleton 23:252–264

Liu L, Shang-Guan K, Zhang B, Liu X, Yan M, Zhang L, Shi Y, Zhang M, Qian Q, Li J, Zhou Y (2013) Brittle Culm1, a COBRA-like protein, functions in cellulose assembly through binding cellulose microfibrils. PLoS Genet 9:e1003704

Liu Z, Schneider R, Kesten C, Zhang Y, Somssich M, Zhang Y, Fernie AR, Persson S (2016) Cellulose-microtubule uncoupling proteins prevent lateral displacement of microtubules during cellulose synthesis in Arabidopsis. Dev Cell 38:305–315

Majda M, Grones P, Sintorn I-M, Vain T, Milani P, Krupinski P, Zagórska-Marek B, Viotti C, Jönsson H, Mellerowicz EJ, Hamant O, Robert S (2017) Mechanochemical polarization of contiguous cell walls shapes plant pavement cells. Dev Cell 43:290–304.e4

Martinière A, Gayral P, Hawes C, Runions J (2011) Building bridges: formin1 of Arabidopsis forms a connection between the cell wall and the actin cytoskeleton. Plant J 66:354–365

Maruta S, Mitsui T, Umezu N, Umeki N, Kondo K (2010) Characterization of a novel rice kinesin O12 with a calponin homology domain. J Biochem 149:91–101

Mathur J, Chua NH (2000) Microtubule stabilization leads to growth reorientation in Arabidopsis trichomes. Plant Cell 12:465–477

Mathur J, Spielhofer P, Kost B, Chua N (1999) The actin cytoskeleton is required to elaborate and maintain spatial patterning during trichome cell morphogenesis in Arabidopsis thaliana. Development 126:5559–5568

Mathur J, Mathur N, Kirik V, Kernebeck B, Purushottam B, Hülskamp M, Srinivas BP, Hulskamp M (2003a) Arabidopsis CROOKED encodes for the smallest subunit of the ARP2/3 complex and controls cell shape by region specific fine F-actin formation. Development 130:3137–3146

Mathur J, Mathur N, Kernebeck B, Hulskamp M, Hülskamp M (2003b) Mutations in actin-related proteins 2 and 3 affect cell shape development in Arabidopsis. Plant Cell 15:1632–1645

McMichael CM, Bednarek SY (2013) Cytoskeletal and membrane dynamics during higher plant cytokinesis. New Phytol 197:1039–1057

Mineyuki Y, Palevitz BA (1990) Relationship between preprophase band organization, F-actin and the division site in Allium. J Cell Sci 97:283-LP-295

Müller S, Jürgens G (2016) Plant cytokinesis-no ring, no constriction but centrifugal construction of the partitioning membrane. Semin Cell Dev Biol 53:10–18

Oda Y (2005) Regulation of secondary cell wall development by cortical microtubules during tracheary element differentiation in Arabidopsis cell suspensions. Plant Physiol 137:1027–1036

Oda Y, Fukuda H (2012) Initiation of cell wall pattern by a rho- and microtubule-driven symmetry breaking. Science 337:1333–1336

Oda Y, Fukuda H (2013a) Rho of plant GTPase signaling regulates the behavior of Arabidopsis kinesin-13A to establish secondary cell wall patterns. Plant Cell 25:4439–4450

Oda Y, Fukuda H (2013b) Spatial organization of xylem cell walls by ROP GTPases and microtubule-associated proteins. Curr Opin Plant Biol 16:743–748

Palevitz BA (1987) Actin in the preprophase band of Allium cepa. J Cell Biol 104:1515–1519

Panteris E (2008) Cortical actin filaments at the division site of mitotic plant cells: a reconsideration of the "actin-depleted zone". New Phytol 179:334–341

Panteris E, Apostolakos P, Galatis B (1993a) Microtubule organization and cell morphogenesis in two semi-lobed cell types of Adiantum capillus-veneris L. leaflets. New Phytol 125:509–520

Panteris E, Apostolakos P, Galatis B (1993b) Microtubules and morphogenesis in ordinary epidermal cells of Vigna sinensis leaves. Protoplasma 174:91–100

Panteris E, Apostolakos P, Galatis B (1994) Sinuous ordinary epidermal cells: behind several patterns of waviness, a common morphogenetic mechanism. New Phytol 127:771–780

Paredez AR, Somerville CR, Ehrhardt DW (2006) Visualization of cellulose synthase demonstrates functional association with microtubules. Science 312:1491–1495

Peremyslov VV, Prokhnevsky AI, Dolja VV (2010) Class XI myosins are required for development, cell expansion, and F-actin organization in Arabidopsis. Plant Cell 22:1883–1897

Pesquet E, Korolev AV, Calder G, Lloyd CW (2010) The microtubule-associated protein AtMAP70-5 regulates secondary wall patterning in Arabidopsis wood cells. Curr Biol 20:744–749

Petrášek J, Schwarzerová K (2009) Actin and microtubule cytoskeleton interactions. Curr Opin Plant Biol 12:728–734

Pratap Sahi V, Cifrová P, García-González J, Kotannal Baby I, Mouillé G, Gineau E, Müller K, Baluška F, Soukup A, Petrášek J, Schwarzerová K (2017) Arabidopsis thaliana plants lacking the ARP2/3 complex show defects in cell wall assembly and auxin distribution. Ann Bot 122:777–789

Preuss ML, Kovar DR, Lee Y-RJ, Staiger CJ, Delmer DP, Liu B (2004) A plant-specific kinesin binds to actin microfilaments and interacts with cortical microtubules in cotton fibers. Plant Physiol 136:3945–3955

Qin T, Liu X, Li J, Sun J, Song L, Mao T (2014) Arabidopsis microtubule-destabilizing protein 25 functions in pollen tube growth by severing actin filaments. Plant Cell 26:325–339

Qiu J-L, Jilk R, Marks MD, Szymanski DB (2002) The Arabidopsis SPIKE1 gene is required for normal cell shape control and tissue development. Plant Cell 14:101–118

Rasmussen CG, Wright AJ, Müller S (2013) The role of the cytoskeleton and associated proteins in determination of the plant cell division plane. Plant J 75:258–269

Roberts AW, Frost AO, Roberts EM, Haigler CH (2004) Roles of microtubules and cellulose microfibril assembly in the localization of secondary-cell-wall deposition in developing tracheary elements. Protoplasma 224:217–229

Rosero A, Žárský V, Cvrčková F (2013) AtFH1 formin mutation affects actin filament and microtubule dynamics in *Arabidopsis thaliana*. J Exp Bot 64:585–597

Rosero A, Oulehlová D, Stillerová L, Schiebertová P, Grunt M, Žárský V, Cvrčková F (2016) Arabidopsis FH1 formin affects cotyledon pavement cell shape by modulating cytoskeleton dynamics. Plant Cell Physiol 57:488–504

Saedler R, Mathur N, Srinivas BP, Kernebeck B, Hulskamp M, Mathur J (2004a) Actin control over microtubules suggested by DISTORTED2 encoding the Arabidopsis ARPC2 subunit homolog. Plant Cell Physiol 45:813–822

Saedler R, Zimmermann I, Mutondo M, Hülskamp M (2004b) The Arabidopsis KLUNKER gene controls cell shape changes and encodes the AtSRA1 homolog. Plant Mol Biol 56:775–782

Sambade A, Findlay K, Schaeffner AR, Lloyd CW, Buschmann H (2014) Actin-dependent and -independent functions of cortical microtubules in the differentiation of Arabidopsis leaf trichomes. Plant Cell 26:1629–1644

Sampathkumar A, Lindeboom JJ, Debolt S, Gutierrez R, Ehrhardt DW, Ketelaar T, Persson S (2011) Live cell imaging reveals structural associations between the actin and microtubule cytoskeleton in Arabidopsis. Plant Cell 23:2302–2313

Sampathkumar A, Gutierrez R, McFarlane HE, Bringmann M, Lindeboom J, Emons AM, Samuels L, Ketelaar T, Ehrhardt DW, Persson S (2013) Patterning and lifetime of plasma membrane-localized cellulose synthase is dependent on actin organization in Arabidopsis interphase cells. Plant Physiol 162:675–688

Sampathkumar A, Krupinski P, Wightman R, Milani P, Berquand A, Boudaoud A, Hamant O, Jonsson H, Meyerowitz EM (2014) Subcellular and supracellular mechanical stress prescribes cytoskeleton behavior in Arabidopsis cotyledon pavement cells. elife 3:e01967

Sanchez-Rodriguez C et al (2012) CHITINASE-LIKE1/POM-POM1 and its homolog CTL2 are glucan-interacting proteins important for cellulose biosynthesis in Arabidopsis. Plant Cell 24:589–607

Sano T, Higaki T, Oda Y, Hayashi T, Hasezawa S (2005) Appearance of actin microfilament 'twin peaks' in mitosis and their function in cell plate formation, as visualized in tobacco BY-2 cells expressing GFP-fimbrin. Plant J 44:595–605

Sapala A et al (2018) Why plants make puzzle cells, and how their shape emerges. Elife 7:1–32

Schaefer E, Belcram K, Uyttewaal M, Duroc Y, Goussot M, Legland D, Laruelle E, de Tauzia-Moreau M-L, Pastuglia M, Bouchez D (2017) The preprophase band of microtubules controls the robustness of division orientation in plants. Science 356:186–189

Schwab B, Mathur J, Saedler RR, Schwarz H, Frey B, Scheidegger C, Hulskamp M (2003) Regulation of cell expansion by the DISTORTED genes in *Arabidopsis thaliana*: actin controls the spatial organization of microtubules. Mol Genet Genomics 269:350–360

Smith LG (2003) Cytoskeletal control of plant cell shape: getting the fine points. Curr Opin Plant Biol 6:63–73

Sotiriou P, Giannoutsou E, Panteris E, Galatis B, Apostolakos P (2018) Local differentiation of cell wall matrix polysaccharides in sinuous pavement cells: its possible involvement in the flexibility of cell shape. Plant Biol (Stuttg) 20:223–237

Spinner L et al (2013) A protein phosphatase 2A complex spatially controls plant cell division. Nat Commun 4:1863

Sugiyama Y, Wakazaki M, Toyooka K, Fukuda H, Oda Y (2017) A novel plasma membrane-anchored protein regulates xylem cell-wall deposition through microtubule-dependent lateral inhibition of rho GTPase domains. Curr Biol 27:2522–2528.e4

Sugiyama Y, Nagashima Y, Wakazaki M, Sato M, Toyooka K, Fukuda H, Oda Y (2019) A rho-actin signaling pathway shapes cell wall boundaries in Arabidopsis xylem vessels. Nat Commun 10:468

Sun T, Li S, Ren H (2017) OsFH15, a class I formin, interacts with microfilaments and microtubules to regulate grain size via affecting cell expansion in rice. Sci Rep 7:6538

Szymanski DB (2009) Plant cells taking shape: new insights into cytoplasmic control. Curr Opin Plant Biol 12:735–744

Szymanski DB, Cosgrove DJ (2009) Dynamic coordination of cytoskeletal and cell wall systems during plant cell morphogenesis. Curr Biol 19:R800–R811

Szymanski DB, Marks MD, Wick SM (1999) Organized F-actin is essential for normal trichome morphogenesis in Arabidopsis. Plant Cell 11:2331–2347

Takeuchi M, Karahara I, Kajimura N, Takaoka A, Murata K, Misaki K, Yonemura S, Staehelin LA, Mineyuki Y (2016) Single microfilaments mediate the early steps of microtubule bundling during preprophase band formation in onion cotyledon epidermal cells. Mol Biol Cell 27:1809–1820

Takeuchi M, Staehelin LA, Mineyuki Y (2017) Actin-microtubule interaction in plants. In: Cytoskeleton - structure, dynamics, function and disease. InTech, Rijeka

Tolmie F, Poulet A, McKenna J, Sassmann S, Graumann K, Deeks M, Runions J (2017) The cell wall of *Arabidopsis thaliana* influences actin network dynamics. J Exp Bot 68:4517–4527

Traas JA, Doonan JH, Rawlins DJ, Shaw PJ, Watts J, Lloyd CW (1987) An actin network is present in the cytoplasm throughout the cell cycle of carrot cells and associates with the dividing nucleus. J Cell Biol 105:387–395

Tseng K-F, Wang P, Lee Y-RJ, Bowen J, Gicking AM, Guo L, Liu B, Qiu W (2018) The preprophase band-associated kinesin-14 OsKCH2 is a processive minus-end-directed microtubule motor. Nat Commun 9:1067

Vain T, Crowell EF, Timpano H, Biot E, Desprez T, Mansoori N, Trindade LM, Pagant S, Robert S, Hofte H, Gonneau M, Vernhettes S (2014) The cellulase KORRIGAN is part of the cellulose synthase complex. Plant Physiol 165:1521–1532

Vandavasi VG et al (2016) A structural study of CESA1 catalytic domain of Arabidopsis cellulose synthesis complex: evidence for CESA Trimers. Plant Physiol 170:123–135

Verger S, Long Y, Boudaoud A, Hamant O (2018) A tension-adhesion feedback loop in plant epidermis. elife 7:1–25

Vőfély RV, Gallagher J, Pisano GD, Bartlett M, Braybrook SA (2019) Of puzzles and pavements: a quantitative exploration of leaf epidermal cell shape. New Phytol 221:540–552

Vosolsobě S, Petrášek J, Schwarzerová K (2017) Evolutionary plasticity of plasma membrane interaction in DREPP family proteins. Biochim Biophys Acta Biomembr 1859:686–697

Walker KL, Müller S, Moss D, Ehrhardt DW, Smith LG (2007) Arabidopsis TANGLED identifies the division plane throughout mitosis and cytokinesis. Curr Biol 17:1827–1836

Wang J, Zhang Y, Wu J, Meng L, Ren H (2013) At FH16, an Arabidopsis type II Formin, binds and bundles both microfilaments and microtubules, and preferentially binds to microtubules. J Integr Plant Biol 55:1002–1015

Welch MD, DePace AH, Verma S, Iwamatsu A, Mitchison TJ (1997) The human Arp2/3 complex is composed of evolutionarily conserved subunits and is localized to cellular regions of dynamic actin filament assembly. J Cell Biol 138:375–384

Welch MD, Rosenblatt J, Skoble J, Portnoy DA, Mitchison TJ (1998) Interaction of human Arp2/3 complex and the *Listeria monocytogenes* ActA protein in actin filament nucleation. Science 281:105–108

Wickstead B, Gull K (2007) Dyneins across eukaryotes: a comparative genomic analysis. Traffic 8:1708–1721

Xu XM, Zhao Q, Rodrigo-Peiris T, Brkljacic J, He CS, Muller S, Meier I (2008) RanGAP1 is a continuous marker of the Arabidopsis cell division plane. Proc Natl Acad Sci 105:18637–18642

Xu T, Wen M, Nagawa S, Fu Y, Chen JG, Wu MJ, Perrot-Rechenmann C, Friml J, Jones AM, Yang Z (2010) Cell surface- and rho GTPase-based Auxin signaling controls cellular Interdigitation in Arabidopsis. Cell 143:99–110

Xu T, Nagawa S, Yang Z (2011) Uniform auxin triggers the rho GTPase-dependent formation of interdigitation patterns in pavement cells. Small GTPases 2:227–232

Xu T et al (2014) Cell surface ABP1-TMK Auxin-sensing complex activates ROP GTPase signaling. Science 343:1025–1028

Yanagisawa M, Desyatova AS, Belteton SA, Mallery EL, Turner JA, Szymanski DB (2015) Patterning mechanisms of cytoskeletal and cell wall systems during leaf trichome morphogenesis. Nat Plants 1:1–8

Yanagisawa M, Alonso JM, Szymanski DB (2018) Microtubule-dependent confinement of a cell signaling and actin polymerization control module regulates polarized cell growth. Curr Biol 28:2459–2466.e4

Zhang XG, Dyachok J, Krishnakumar S, Smith LG, Oppenheimer DG (2005) IRREGULAR TRICHOME BRANCH1 in Arabidopsis encodes a plant homolog of the actin-related protein2/3 complex activator Scar/WAVE that regulates actin and microtubule organization. Plant Cell 17:2314–2326

Zhang Z, Zhang Y, Tan H, Wang Y, Li G, Liang W, Yuan Z, Hu J, Ren H, Zhang D (2011a) RICE MORPHOLOGY DETERMINANT encodes the type II Formin FH5 and regulates rice morphogenesis. Plant Cell 23:681–700

Zhang C, Halsey LE, Szymanski DB (2011b) The development and geometry of shape change in *Arabidopsis thaliana* cotyledon pavement cells. BMC Plant Biol 11:27

Zhang C, Mallery EL, Szymanski D (2013) ARP2/3 localization in Arabidopsis leaf pavement cells: a diversity of intracellular pools and cytoskeletal interactions. Front Plant Sci 4:238

Zhang W, Cai C, Staiger CJ (2019) Myosins XI are involved in exocytosis of cellulose synthase complexes. Plant Physiol 179(4):1537–1555. https://doi.org/10.1104/pp.19.00018

Zhu C, Ganguly A, Baskin TI, McClosky DD, Anderson CT, Foster C, Meunier KA, Okamoto R, Berg H, Dixit R (2015) The fragile Fiber1 kinesin contributes to cortical microtubule-mediated trafficking of cell wall components. Plant Physiol 167:780–792

Chapter 2
Insights into the Cell Wall and Cytoskeletal Regulation by Mechanical Forces in Plants

Yang Wang, Ritika Kulshreshtha, and Arun Sampathkumar

Abstract Morphogenesis is a highly controlled biological process that causes a plant to develop particularly shaped organs. During this process, directional growth of cells is achieved by a combinatorial action of isotropic turgor driven expansion, which is spatially constrained or relaxed by either deposition or modification of cell wall polymers. Immense networks of genes and signaling cascades have been identified to govern the process of morphogenesis. However, for shape changes to occur concurrent modulations to structural properties of the cell wall that encapsulates plant cells are necessary. The microtubule cytoskeleton via its regulation of cellulose deposition and the activity of cell wall modifying enzymes controls directional growth and cell wall stiffness, respectively influencing morphogenesis. In this chapter, we outline the components that contribute to the mechanics of plant cells and organs that ultimately regulate plant growth and form.

2.1 Plant Cell Wall

All plant cells are encased by an exoskeleton called the cell wall. Plant cell walls are usually divided into two types: primary cell wall and secondary cell wall. Primary cell walls are extensible and present in all growing cells. Whereas secondary cell walls are rigid and durable and only occur in cells of tissues that need high mechanical support such as plant stem and vessels. Secondary walls are deposited interior of primary walls after cell stops growing and usually much thicker than primary cell walls. The primary cell wall is composed of cellulose, hemicellulose, pectin, and small amounts of glycosylated proteins. In comparison, the secondary cell wall contains a large amount of cellulose, hemicellulose, lignin, and glycoproteins. Pectin is nearly absent in secondary cell wall, and the polysaccharide ratio of hemicellulose is also different from the primary cell wall (Ivakov and Persson 2012). Due to the mechanical property difference, the primary cell wall is cell growth related while the secondary

Y. Wang · R. Kulshreshtha · A. Sampathkumar (✉)
Max Planck Institute of Molecular Plant Physiology, Potsdam, Germany
e-mail: Sampathkumar@mpimp-golm.mpg.de

© Springer Nature Switzerland AG 2019
V. P. Sahi, F. Baluška (eds.), *The Cytoskeleton*, Plant Cell Monographs 24,
https://doi.org/10.1007/978-3-030-33528-1_2

wall is more involved in providing physical strength and water transport throughout the plant. Apart from this, the plant cell wall plays several important roles, here we will focus only on the mechanical aspects of primary cell walls.

2.1.1 Cellulose

Cellulose is composed of sheets of hydrogen-bonded polysaccharide chains made up of β $(1 \rightarrow 4)$ glycosidic linked D-glucose units. As the major tension bearing element in the cell wall, cellulose microfibrils constitute the most abundant polymer of the plant cell wall, cross-linked by hemicellulose and pectin polymers. Primary cell wall cellulose microfibril is composed of 12–36 glucan chains with a diameter of a few nanometers (Fernandes et al. 2011; Kubicki et al. 2018; Thomas et al. 2013). Cellulose microfibrils are synthesized at the plasma membrane by cellulose synthase complex (CSC), which is a rosette-like protein complex consisting of different cellulose synthase (CESA) proteins (McFarlane et al. 2014, Bidhendi and Geitmann). CSC is assembled at the Golgi apparatus and secreted to the plasma membrane. Among the 10 CESA proteins in *Arabidopsis*, CESA1, CESA3, and CESA6, as well as CESA6 related CESA2, CESA5, CESA9, participate in the synthesis of primary cell wall cellulose microfibril (Desprez et al. 2007; McFarlane et al. 2014; Ambrose et al. 2011). Null mutants of *CESA1* and *CESA3* show gametophytic lethal phenotype, while *CESA6* mutant (*procuste, prc1-1*) displays a relatively stunted phenotype due to its functional redundancy with the CESA6 related synthases (Desprez et al. 2007; Persson et al. 2007). Disruption of cellulose was shown to reduce the mechanical stiffness of plant cell walls (Sampathkumar et al. 2019).

2.1.2 Pectin

Pectins are important structural polysaccharides contributing to one-third of the cell wall dry mass (Caffall and Mohnen 2009). In contrast to cellulose that is synthesized at the plasma membrane, pectin is assembled inside the Golgi and then secreted to the apoplast by vesicle trafficking and fusion of the Golgi vesicles with the plasma membrane (Driouich et al. 2012). Their main component is homogalacturonan (HG), which is secreted in a highly methyl-esterified form. Enzymes like PECTIN METHYLESTERASE (PME) de-methyl-esterify HG and form load-bearing Ca^{2+}-pectate cross-linked complexes (Hocq et al. 2017). These cross-links affect cell wall porosity and hence the accessibility for several cell wall-remodeling proteins, such as PMEs and EXPANSINS (Cosgrove 2016). Interestingly, recent experiments using *PME5* overexpression showed global cell wall loosening and emergence of ectopic primordia from the SAM of *Arabidopsis* (Peaucelle et al. 2012). This is in contrary to the cell wall stiffening properties of PME (Bidhendi and Geitmann 2016; Parre and Geitmann 2005). PMEs are also under control of their inhibitors, PECTIN

METHYLESTERASE INHIBITOR (PMEi), and their balance is thought to define another level of regulation of cell wall stiffness. *PMEi3* overexpression results in global cell wall stiffening and complete inhibition of organogenesis at the shoot apical meristem (Peaucelle et al. 2011, Braybrook and Peaucelle 2013).

2.1.3   Hemicellulose

Hemicellulose is considered as a group of polysaccharides in plant cell walls that have β-(1→4)-linked backbones with an equatorial configuration, including xyloglucans, xylans, mannans, glucomannans, and β-(1→3, 1→4)-glucans (Scheller and Ulvskov 2010). In primary cell walls of dicot plants, Xyloglucan (XyG) is the most abundant hemicellulose saccharide. Several XYLOGLUCAN XYLOSYLTRANSFERASES (XXTs) are enzymes that are involved in the synthesis of the XyGs. XXTs are transmembrane proteins present in the Golgi membrane with the catalytic domain facing the Golgi lumen. *xxt1 xxt2* double mutants have a complete lack of XyGs exhibit minor morphological phenotypes, yet the plants are viable (Cavalier et al. 2008; Xiao et al. 2016). Hemicelluloses are known to interact with cellulose microfibrils directly through hydrogen bond which might influence the mechanical properties of the cell wall (McFarlane et al. 2014; Scheller and Ulvskov 2010). The combined loss of XXT1 XXT2 was also shown to influence the mechanics of the cell wall (Cavalier et al. 2008; Park and Cosgrove 2012) suggesting that they might play an important role in cross-linking cellulose microfi brils. However, the direct contribution of XyG to the mechanics of the cell wall is minimal (Cosgrove 2016) as enzymatic disruption of XyG did not cause any changes to the physical properties of the cell wall (Saladie et al. 2006).

2.1.4   Expansins

Apart from the polysaccharides, there are also a small number of proteins some of which are detailed above, most of them are glycosylated. Among these cell wall proteins, expansins that are classified into α-expansins and β-expansins based on sequence phylogeny are indicated to play an important role in cell expansion. In plants, α-expansins are known to modulate wall loosening nonenzymatically in a pH-dependent fashion (Cosgrove 2015). Studies based on bacterial expansins suggested that they bind both to cellulose and pectin (Wang et al. 2013). Such binding is also proposed to occur at biomechanical hotspots that are present between closely interacting cellulose microfirbrils (Park and Cosgrove 2015). This was shown to be the case by means of modified solid-state Nuclear Magnetic Resonance spectroscopy based study in which expansin binding sites to cellulose was also enriched with XyG and have a cellulose microfibril conformation that is different from the major proportion of cellulose present in the cell wall (Wang et al. 2013).

2.2 Morphogenesis in Plants

2.2.1 *Plant Cell Growth*

During growth, plant cells accumulate organic and inorganic salts in the cytosol mediated by plasma membrane-localized ATPases that act as molecular pumps. Such changes in ion concentration result in the passage of water into or out of cells by means of osmosis until an equilibrium is reached. The buildup of water results in accumulation of hydrostatic pressure also termed as turgor pressure in the cell (Fricke 2017). Turgor pressure in plant cells is several folds higher than atmospheric pressure reaching up to 2 MPa, providing structural integrity to plant cells. To counter such high turgor pressures plant cells are encased by a cell wall, which is a tough yet malleable structure. While counterbalancing turgor pressure, plant cell wall is also experiencing in plane mechanical tension. Such stresses result in elastic (reversible) deformation of the cell wall until they exceed the elastic limit or yield point after which plastic (irreversible) deformation occurs. The cell wall when under sustained physical stress as opposed to rapid and short burst of increased stress results in viscoelastic or viscoplastic deformation that is time dependent. Coupled to these merely mechanical feedbacks, growth of plant cell is accomplished by cell wall biochemical remodeling and biosynthesis of new cell wall components (Cosgrove 2018).

Structural properties of the cell wall apart by constraining and yielding to physical stress also influence growth directionality. Cellulose microfibrils being the stiffest polymer (approximately 100 GPa load-bearing ability) in cell wall play an important role in this process. Several experimental observations showed that the newly deposited cellulose microfibrils at the inner surface of cell wall were transversely oriented in cylindrical cells (Eng and Sampathkumar 2018; Anderson et al. 2010; Refrégier et al. 2004). This suggests that isotropic (nondirectional) turgor pressure results in anisotropic (directional) expansion along an axis perpendicular to the major orientation of cellulose microfibrils. Supportive of this, cellulose deficient mutants and plants treated with cellulose synthesis inhibitor showed phenotypes of radial swelling and reduced elongation in hypocotyls and roots (Refrégier et al. 2004; Arioli et al. 1998; Fagard et al. 2000; Fujita et al. 2013; Sugimoto et al. 2001). Research on trichomes, which is undergoing fast anisotropic tip growth, showed that in rapidly growing regions the cell wall was thinner than the walls transverse to growth axis (Yanagisawa et al. 2015). Therefore, cellulose microfibril alignment is an important element for the regulation of anisotropic cell expansion.

2.2.2 *Microtubule in Cellulose Biosynthesis and Directional Growth*

What regulates cellulose microfibril organization in plant cells? Microtubule cytoskeleton in hypocotyl cells hyperalign perpendicularly to the major growth axis, which is consistent with the newly formed microfibril orientation (Green 1962). These observations lead to the postulation of a hypothesis whereby cellulose microfibril orientation could be dependent on the underlying microtubule cytoskeleton network. Several studies including spinning disk confocal microscope-based time lapse imaging has convincingly showed that CSC migrates bidirectionally along cortical microtubules at the plasma membrane (Paredez et al. 2006). In addition to this, cortical microtubule acts to position delivery of CSC to the plasma membrane and interact with CSC trafficking compartments (Gutierrez et al. 2009). Besides, partial depolymerization of microtubules using oryzalin could lead to an abnormal CSC trajectory disrupting microfibril orientation (Chan et al. 2010; Paredez et al. 2006), resulting in swelling of cells (Anthony and Hussey 1999; Chan et al. 2010). Mutants of microtubule regulatory proteins such as KATANIN and CLASP showed abnormal microtubule organization and swelled cells (Ambrose et al. 2007; Burk et al. 2001). More recently, POM2/CELLULOSE SYNTHASE INTERACTING1 (CSI1) has been identified as a linker protein between CSCs and cortical microtubules, which has direct interaction with microtubules and colocalizes with CSCs. Absence of POM2/CSI1 affects the coalignment of microtubule and CSC movement, CSC velocities and cell elongation defects (Bringmann et al. 2012; Gu et al. 2010; Li et al. 2012; Mei et al. 2012). Taken together, cortical microtubule can serve as tracks guiding CSC movement during cellulose microfibril biosynthesis at the plasma membrane. However, complete removal of microtubules resulted in organized CSC trajectories in hypocotyl cells during growth (Paredez et al. 2006), suggesting other factors could influence CSC trajectories (McFarlane et al. 2014).

2.2.3 Actin Filaments in Cellulose Biosynthesis and Directional Growth

Actin filament, or F-actin, is polymerized from G-actin monomers, is the other component of the plant cytoskeleton machinery. Actin plays an important role in the trafficking of cellular compartments in plants cell. Research has demonstrated that actin mediates deposition of cellulose-related polymers and pectin in tip-growing cells, such as pollen tubes and root hairs (Chen et al. 2007; Hu et al. 2003). It has been shown that the organization of actin filament is responsible for the trafficking of CSC-contained vesicles and the global distribution of CESAs at the plasma membrane (Sampathkumar et al. 2013). Mutants of *actin* and plants treated with actin-depolymerizing drug were observed to display impaired anisotropic cell expansion in root and broadened root hair tips (Baluska et al. 2001; Kandasamy et al. 2009). Actin accumulates at sites of rapid expansion such as tips of pollen tubes (Chen et al. 2007) and apical region, tip of the protonema, of the moss *Physcomitrella patens* (Wu and Bezanilla 2018), presumably trafficking cell wall-related polymers and other proteins to promote growth.

2.2.4   Auxin in Pattern Formation

It is long known that the plant hormone auxin is a key regulator of growth, in particular, cell division, cell expansion, plant tropism, shoot architecture, and lateral organ formation (Heisler et al. 2005; Leyser 2018). Exogenously applied auxin causes acidification of the cell wall, a process that in effect weakens the wall, thereby allowing cell expansion. Auxin exists in a neutral uncharged form when present in the cell wall and is capable of diffusing across the plasma membrane into the cell where it is rendered into a charged ionic state (Rayle and Cleland 1992; Rubery and Sheldrake 1974). The presence of a polar auxin efflux carrier protein PIN1 at the plasma membrane actively transports auxin across the membrane into the cell wall at the expense of ATP hydrolysis (Rubery and Sheldrake 1974). PIN1 is preferentially localized to the plasma membrane facing the cell with a higher auxin concentration (Heisler et al. 2005). This distribution pattern enables auxin to move against its concentration gradient. This leads to hot spots of high auxin concentration, which marks the sites of new organ formation (Taylor-Teeples et al. 2016). Atomic force microscopy measured tissue rigidity at the shoot apex shows that upon auxin application, a reduction in elastic modulus was observed prior to organ initiation (Braybrook and Peaucelle 2013; Kierzkowski et al. 2012). Further experimental data revealed that auxin can modify the wall rigidity through inducing local de-methyl-esterification of HG in subepidermal tissues (Braybrook and Peaucelle 2013), allowing the outgrowth of emerging organs.

2.3   Biomechanics Behind Plant Morphogenesis

Mechanical forces such as tension, compression, and shear stress are omnipresent and critically important for the growth and development of multicellular organisms. Plant cells are under constant mechanical tension due to its internal turgor pressure as well as due to the growth of neighboring cells. Interdisciplinary approaches that combine finite element model, measurement of mechanical properties using micro- or nanoscale indentation experiments, and molecular approaches that involve live-cell imaging and mutant analysis have started to reveal the importance of mechanics in plant growth and morphogenesis (Milani et al. 2013; Routier-Kierzkowska and Smith 2013). We can summarize the above aspects on cell wall mechanics in the following schematic which we call the mechanical feedback loop (Fig. 2.1) (Sampathkumar et al. 2014b): Directional mechanical stress arises due to the intrinsic turgor pressure in combination with cell or tissue geometry (Hamant et al. 2008; Sampathkumar et al. 2014a). The microtubule cytoskeleton aligns along the principal direction of anisotropic stress, which leads to local anisotropic expansion via the cellulose synthesis machinery. Mechanical forces also influence auxin gradients, which further amplifies microtubule isotropy in addition to acidification of local cell walls regions leading to chemical modifications of wall stiffness (Sassi et al. 2014). These patterns together with the turgor pressure create spatially varying local growth rates, leading to shape changes of a cell or tissue. This

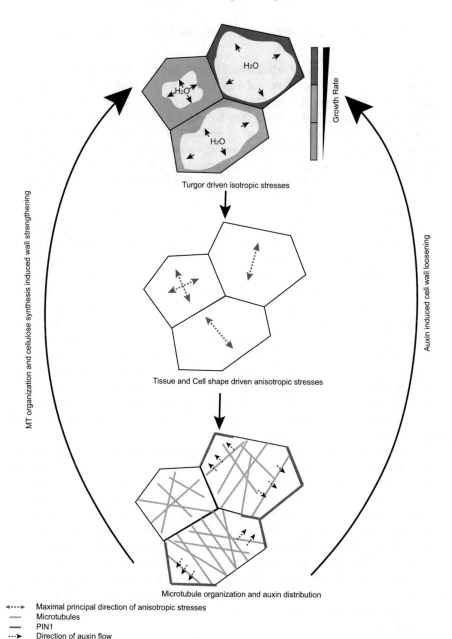

Fig. 2.1 Mechanical feedback loop in cell growth

growth results in the generation of new mechanical signals because all cells in a tissue respond as an ensemble and are all held together via their cell walls. This mechanical feedback loop reconstitutes the role of physical stresses in forming cell and tissue shapes.

2.3.1   Mechanical Patterns Affect Cytoskeleton Alignment

Recent advances in imaging technologies in combination with finite element models show a strong correlation between the alignment of microtubules and stress sensors arising due to the tissue and cellular geometry. At the subcellular level, in epidermal pavement cells of cotyledon tissue, microtubules are highly ordered at the indenting domains of the cell which is a region predicted to experience anisotropic stress (Fig. 2.2). Atomic force microscopy based measurement of cell wall stiffness in these cells shows stiffer fibril-like structures present in the indenting domains consistent with the hypothesis that cellulose microfibrils reinforce regions that are experiencing higher mechanical stress (Sampathkumar et al. 2014a). Also, research in *Arabidopsis* shoot apical meristem (SAM), a dome-shaped structure, found the apex region showed supracellular coalignment between microtubules and principal stress direction (Fig. 2.2). The central domain of the SAM contains isotropically ordered microtubules, whereas the peripheral domain possesses a dominant circumferential orientation of microtubules concomitant with identically ordered stress patterns. The boundary domains between SAM and emerging organs experience highly anisotropic stress with a preferential alignment along with the cell's long axis matching the patterns of microtubule organization (Hamant et al. 2008). Further micro-mechanical manipulation experiments such as direct compression and laser ablation demonstrated that changes in stress tensors resulted in concurrent changes in microtubule organization (Fig. 2.2) (Hamant et al. 2008; Sampathkumar et al. 2014a). In the reduced function mutant of microtubule-severing protein KATANIN, microtubules do not efficiently respond to changes in mechanical forces, suggesting microtubule severing plays an important role in microtubule response to mechanical perturbation (Uyttewaal et al. 2012).

While the role of microtubules in perceiving mechanical changes is well demonstrated, the influence on actin in such a process is less well understood. Studies have shown actin filaments rapidly aggregate at the site of mechanical contact with microneedle in both *Arabidopsis* cotyledon cells and hypocotyl cells (Branco et al. 2017; Hardham et al. 2008). Nanoindentation-based quantification revealed that in *Arabidopsis* hypocotyl cells, actin reorganization can be triggered with a force as small as 4 μN only for 21.6 s (Branco et al. 2017). Whereas, microtubule array aggregation occurred around 6.5 h after application of ~2 mN indentation force in the SAM (Louveaux et al. 2016b). It could occur partly due to the polymerization speed of actin filaments being much faster than microtubules (Eng and Sampathkumar 2018). The role and detailed dynamics of actin upon mechanical perception and response are less well understood.

2.3.2   Mechanical Pattern Affects Cell Division Plane Orientation

Another important feature that is demonstrated to be under the influence of mechanics is the choice of division plane orientation in plant cells. Sachs proposed that cells

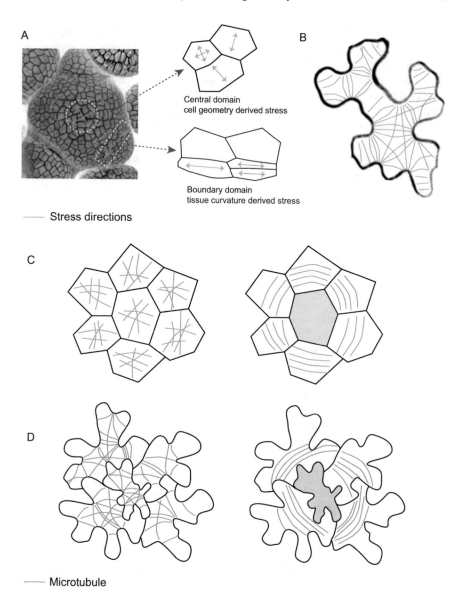

Fig. 2.2 Stress distribution patterns and Microtubule (MT) organization (**a**) Stress distribution pattern in *Arabidopsis* shoot apical meristem (SAM). Orange arrows depict the maximal principal stress direction. Isotropic stresses are observed in the central region, circumferential stress in the peripheral region, and anisotropic stress along the boundary region. In the homogeneously growing center region, the stress pattern within single cell originates from the cell geometry, whereas in boundary regions, the single cell stress pattern is usually biased by the tissue curvature or heterogeneous growth of neighboring cells. (**b**) Mechanical pattern in *Arabidopsis* pavement cell. Orange lines show stress patterns in pavement cells. Stresses in indented regions are more anisotropic. (**c**) MT organization in SAM center region cells and realignment upon mechanical perturbation by cell ablation. (**d**) MT organization in pavement cells and realignment upon mechanical perturbation by cell ablation. Gray regions indicate ablated cells; green lines represent MT organization

divide symmetrically by the formation of the new cell walls at a right angle to the existing walls (Sachs 1878). Later, Hofmeister proposed that the newly formed cell wall is perpendicular to the major growth axis (Hofmeister 1863). Then Errera's rule proposes that cells divide along the shortest path across the center to minimize the area of the new cell wall (Errera 1888). These rules emphasized the importance of cell geometry on cell division plane orientation. However, deviations from these rules have been observed in plant cells. For example, in *Arabidopsis* SAM, Louveaux et al. found that although cell divisions in the center region follow the Errera's rule, the divisions in the highly curved saddle-shaped boundary region of SAM and an emerging organ occurs along the long axis of the cell (Louveaux et al. 2016a). This is consistent with highly anisotropic stresses along the crease in the boundary region (Hamant et al. 2008). This suggests that mechanical forces could play an important role in regulating cell division plane orientation. Supportive of this the newly formed walls around the site of laser ablation in the central domain of SAM where circumferentially aligned similar to the predicted alignment of stress after mechanical perturbation at the SAM ignoring cell geometry-based cues. Thus, it is proposed that cell divides along the maximum direction of stress (Louveaux et al. 2016a).

Cytoskeleton dynamics is an essential feature that influences division plane orientation. It is now well-known that, in late interphase, cortical microtubules form a ring-like structure at the equatorial plane of the cell termed preprophase band (PPB) that predicts cell division plane orientation. In addition to microtubules, actin filaments were also observed in the PPB (Livanos and Muller 2019; Palevitz 1987; Vanstraelen et al. 2006). Before cell division, cytoplasmic strands containing actin and microtubules position nucleus at the center of the cell (Flanders et al. 1990; Goodbody et al. 1991), and then coalesce into a plane named phragmosome, whose position coincides with PPB. After the disappearance of PPB, transient cortical actin depleted zone marks the future division plane (Smith 2001). Given the knowledge that cortical microtubule and actin arrangement could be influenced by mechanical pattern, mechanical perception and response by the actin cytoskeleton could be the key factors in regulating division plane orientation (Louveaux and Hamant 2013).

2.3.3 Auxin Distribution Is Under Tight Mechanical Control

Mechanical perturbation experiments proved that similar to microtubule alignment, auxin efflux carrier concentrates to the most mechanically stressed cell wall (Heisler et al. 2010), thus resulting in a high concentration of auxin in the mechanically stressed region in tissue. By inducing HG de-methyl-esterification, auxin loosens the stressed walls and enhances the rapid expansion of these walls along with the increase of turgor pressure (Braybrook and Peaucelle 2013), possibly in turn concentrates more auxin transport to these tissues. Furthermore, actin filaments have been shown to be necessary for PIN1 polar localization (Geldner et al. 2001). An exploration into mechanical regulation of actin filaments and auxin efflux carrier could possibly make a more complete picture about mechanics in plant morphogenesis.

2.4 Conclusion

Over the last two decades, several findings have contributed toward our understanding of the mechanics behind plant growth and development. This has been fueled by the application of quantitative approaches that provide information about behavior or proteins at a subcellular scale and how such properties contribute to the growth dynamics and mechanics at a cellular and organ level. In addition to this, the development of finite element models of growing plant cells and tissues has allowed us to test and propose new hypotheses on morphogenesis. What we lack now is the integration of well-known gene regulatory networks in such a biomechanical feedback loop as well as the identification of sensors of mechanical forces in plants. Efforts are underway in several laboratories to make advances on this topic that would enrich us with knowledge on how plants grow.

References

Ambrose JC, Shoji T, Kotzer AM, Pighin JA, Wasteneys GO (2007) The Arabidopsis CLASP gene encodes a microtubule-associated protein involved in cell expansion and division. Plant Cell 19:2763–2775

Ambrose C, Allard JF, Cytrynbaum EN, Wasteneys GO (2011) A CLASP-modulated cell edge barrier mechanism drives cell-wide cortical microtubule organization in Arabidopsis. Nat Commun 2:430

Anderson CT, Carroll A, Akhmetova L, Somerville C (2010) Real-time imaging of cellulose reorientation during cell wall expansion in Arabidopsis roots. Plant Physiol 152:787–796

Anthony RG, Hussey PJ (1999) Dinitroaniline herbicide resistance and the microtubule cytoskeleton. Trends Plant Sci 4:112–116

Arioli T, Peng L, Betzner AS, Burn J, Wittke W, Herth W, Camilleri C, Hofte H, Plazinski J, Birch R, Cork A, Glover J, Redmond J, Williamson RE (1998) Molecular analysis of cellulose biosynthesis in Arabidopsis. Science 279:717–720

Baluska F, Jasik J, Edelmann HG, Salajova T, Volkmann D (2001) Latrunculin B-induced plant dwarfism: plant cell elongation is F-actin-dependent. Dev Biol 231:113–124

Bidhendi AJ, Geitmann A (2016) Relating the mechanics of the primary plant cell wall to morphogenesis. J Exp Bot 67:449–461

Branco R, Pearsall EJ, Rundle CA, White RG, Bradby JE, Hardham AR (2017) Quantifying the plant actin cytoskeleton response to applied pressure using nanoindentation. Protoplasma 254:1127–1137

Braybrook SA, Peaucelle A (2013) Mechano-chemical aspects of organ formation in *Arabidopsis thaliana*: the relationship between auxin and pectin. PLoS One 8:e57813

Bringmann M, LI E, Sampathkumar A, Kocabek T, Hauser MT, Persson S (2012) POM-POM2/cellulose synthase interacting1 is essential for the functional association of cellulose synthase and microtubules in Arabidopsis. Plant Cell 24:163–177

Burk DH, Liu B, Zhong R, Morrison WH, Ye ZH (2001) A katanin-like protein regulates normal cell wall biosynthesis and cell elongation. Plant Cell 13:807–827

Caffall KH, Mohnen D (2009) The structure, function, and biosynthesis of plant cell wall pectic polysaccharides. Carbohydr Res 344:1879–1900

Cavalier DM, Lerouxel O, Neumetzler L, Yamauchi K, Reinecke A, Freshour G, Zabotina OA, Hahn MG, Burgert I, Pauly M, Raikhel NV, Keegstra K (2008) Disrupting two *Arabidopsis thaliana* xylosyltransferase genes results in plants deficient in xyloglucan, a major primary cell wall component. Plant Cell 20:1519–1537

Chan J, Crowell E, Eder M, Calder G, Bunnewell S, Findlay K, Vernhettes S, Hofte H, Lloyd C (2010) The rotation of cellulose synthase trajectories is microtubule dependent and influences the texture of epidermal cell walls in Arabidopsis hypocotyls. J Cell Sci 123:3490–3495

Chen T, Teng N, Wu X, Wang Y, Tang W, Samaj J, Baluska F, Lin J (2007) Disruption of actin filaments by latrunculin B affects cell wall construction in *Picea meyeri* pollen tube by disturbing vesicle trafficking. Plant Cell Physiol 48:19–30

Cosgrove DJ (2015) Plant expansins: diversity and interactions with plant cell walls. Curr Opin Plant Biol 25:162–172

Cosgrove DJ (2016) Catalysts of plant cell wall loosening. F1000Res 5

Cosgrove DJ (2018) Diffuse growth of plant cell walls. Plant Physiol 176:16–27

Desprez T, Juraniec M, Crowell EF, Jouy H, Pochylova Z, Parcy F, Hofte H, Gonneau M, Vernhettes S (2007) Organization of cellulose synthase complexes involved in primary cell wall synthesis in *Arabidopsis thaliana*. Proc Natl Acad Sci U S A 104:15572–15577

Driouich A, Follet-Gueye ML, Bernard S, Kousar S, Chevalier L, Vicre-Gibouin M, Lerouxel O (2012) Golgi-mediated synthesis and secretion of matrix polysaccharides of the primary cell wall of higher plants. Front Plant Sci 3:79

Eng RC, Sampathkumar A (2018) Getting into shape: the mechanics behind plant morphogenesis. Curr Opin Plant Biol 46:25–31

Errera L (1888) Uber zellformen und seifenblasen. Bot Centralbl 34:395–398

Fagard M, Desnos T, Desprez T, Goubet F, Refregier G, Mouille G, Mccann M, Rayon C, Vernhettes S, Hofte H (2000) PROCUSTE1 encodes a cellulose synthase required for normal cell elongation specifically in roots and dark-grown hypocotyls of Arabidopsis. Plant Cell 12:2409–2424

Fernandes AN, Thomas LH, Altaner CM, Callow P, Forsyth VT, Apperley DC, Kennedy CJ, Jarvis MC (2011) Nanostructure of cellulose microfibrils in spruce wood. Proc Natl Acad Sci U S A 108:E1195–E1203

Flanders DJ, Rawlins DJ, Shaw PJ, Lloyd CW (1990) Nucleus-associated microtubules help determine the division plane of plant epidermal cells: avoidance of four-way junctions and the role of cell geometry. J Cell Biol 110:1111–1122

Fricke W (2017) Turgor pressure. In: eLS. Wiley, Chichester

Fujita M, Himmelspach R, Ward J, Whittington A, Hasenbein N, Liu C, Truong TT, Galway ME, Mansfield SD, Hocart CH, Wasteneys GO (2013) The anisotropy1 D604N mutation in the Arabidopsis cellulose synthase1 catalytic domain reduces cell wall crystallinity and the velocity of cellulose synthase complexes. Plant Physiol 162:74–85

Geldner N, Friml J, Stierhof YD, Jurgens G, Palme K (2001) Auxin transport inhibitors block PIN1 cycling and vesicle trafficking. Nature 413:425–428

Goodbody KC, Venverloo CJ, Lloyd CW (1991) Laser microsurgery demonstrates that cytoplasmic strands anchoring the nucleus across the vacuole of premitotic plant cells are under tension. Implications for division plane alignment. Development 113:931–939

Green PB (1962) Mechanism for plant cellular morphogenesis. Science 138:1404–1405

Gu Y, Kaplinsky N, Bringmann M, Cobb A, Carroll A, Sampathkumar A, Baskin TI, Persson S, Somerville CR (2010) Identification of a cellulose synthase-associated protein required for cellulose biosynthesis. Proc Natl Acad Sci U S A 107:12866–12871

Gutierrez R, Lindeboom JJ, Paredez AR, Emons AM, Ehrhardt DW (2009) Arabidopsis cortical microtubules position cellulose synthase delivery to the plasma membrane and interact with cellulose synthase trafficking compartments. Nat Cell Biol 11:797–806

Hamant O, Heisler MG, Jonsson H, Krupinski P, Uyttewaal M, Bokov P, Corson F, Sahlin P, Boudaoud A, Meyerowitz EM, Couder Y, Traas J (2008) Developmental patterning by mechanical signals in Arabidopsis. Science 322:1650–1655

Hardham AR, Takemoto D, White RG (2008) Rapid and dynamic subcellular reorganization following mechanical stimulation of Arabidopsis epidermal cells mimics responses to fungal and oomycete attack. BMC Plant Biol 8:63

Heisler MG, Ohno C, Das P, Sieber P, Reddy GV, Long JA, Meyerowitz EM (2005) Patterns of auxin transport and gene expression during primordium development revealed by live imaging of the Arabidopsis inflorescence meristem. Curr Biol 15:1899–1911

Heisler MG, Hamant O, Krupinski P, Uyttewaal M, Ohno C, Jonsson H, Traas J, Meyerowitz EM (2010) Alignment between PIN1 polarity and microtubule orientation in the shoot apical meristem reveals a tight coupling between morphogenesis and auxin transport. PLoS Biol 8:e1000516

Hocq L, Pelloux J, Lefebvre V (2017) Connecting homogalacturonan-type pectin remodeling to acid growth. Trends Plant Sci 22:20–29

Hofmeister W (1863) Zusatze und Berichtigungen zu den 1851 veroffentlichen Untersuchungen der Entwicklung hoherer Kryptogamen. Jahrb Wiss Bot 3:259–193

Hu Y, Zhong RQ, Morrison WH, Ye ZH (2003) The Arabidopsis RHD3 gene is required for cell wall biosynthesis and actin organization. Planta 217:912–921

Ivakov A, Persson S (2012) Plant cell walls. In: eLS. Wiley, Chichester

Kandasamy MK, Mckinney EC, Meagher RB (2009) A single vegetative actin isovariant overexpressed under the control of multiple regulatory sequences is sufficient for normal Arabidopsis development. Plant Cell 21:701–718

Kierzkowski D, Nakayama N, Routier-Kierzkowska AL, Weber A, Bayer E, Schorderet M, Reinhardt D, Kuhlemeier C, Smith RS (2012) Elastic domains regulate growth and organogenesis in the plant shoot apical meristem. Science 335:1096–1099

Kubicki JD, Yang H, Sawada D, O'Neill H, Oehme D, Cosgrove D (2018) The shape of native plant cellulose microfibrils. Sci Rep 8:13983

Leyser O (2018) Auxin signaling. Plant Physiol 176:465–479

Li S, Lei L, Somerville CR, Gu Y (2012) Cellulose synthase interactive protein 1 (CSI1) links microtubules and cellulose synthase complexes. Proc Natl Acad Sci U S A 109:185–190

Livanos P, Muller S (2019) Division plane establishment and cytokinesis. Annu Rev Plant Biol 70:239–267

Louveaux M, Hamant O (2013) The mechanics behind cell division. Curr Opin Plant Biol 16:774–779

Louveaux M, Julien JD, Mirabet V, Boudaoud A, Hamant O (2016a) Cell division plane orientation based on tensile stress in *Arabidopsis thaliana*. Proc Natl Acad Sci U S A 113:E4294–E4303

Louveaux M, Rochette S, Beauzamy L, Boudaoud A, Hamant O (2016b) The impact of mechanical compression on cortical microtubules in Arabidopsis: a quantitative pipeline. Plant J 88:328–342

Mcfarlane HE, Doring A, Persson S (2014) The cell biology of cellulose synthesis. Annu Rev Plant Biol 65:69–94

Mei Y, Gao HB, Yuan M, Xue HW (2012) The Arabidopsis ARCP protein, CSI1, which is required for microtubule stability, is necessary for root and anther development. Plant Cell 24:1066–1080

Milani P, Braybrook SA, Boudaoud A (2013) Shrinking the hammer: micromechanical approaches to morphogenesis. J Exp Bot 64:4651–4662

Palevitz BA (1987) Actin in the preprophase band of *Allium cepa*. J Cell Biol 104:1515–1519

Paredez AR, Somerville CR, Ehrhardt DW (2006) Visualization of cellulose synthase demonstrates functional association with microtubules. Science 312:1491–1495

Park YB, Cosgrove DJ (2012) Changes in cell wall biomechanical properties in the xyloglucan-deficient xxt1/xxt2 mutant of Arabidopsis. Plant Physiol 158:465–475

Park YB, Cosgrove DJ (2015) Xyloglucan and its interactions with other components of the growing cell wall. Plant Cell Physiol 56:180–194

Parre E, Geitmann A (2005) Pectin and the role of the physical properties of the cell wall in pollen tube growth of *Solanum chacoense*. Planta 220:582–592

Peaucelle A, Braybrook SA, Le Guillou L, Bron E, Kuhlemeier C, Hofte H (2011) Pectin-induced changes in cell wall mechanics underlie organ initiation in Arabidopsis. Curr Biol 21:1720–1726

Peaucelle A, Braybrook S, Hofte H (2012) Cell wall mechanics and growth control in plants: the role of pectins revisited. Front Plant Sci 3:121

Persson S, Paredez A, Carroll A, Palsdottir H, Doblin M, Poindexter P, Khitrov N, Auer M, Somerville CR (2007) Genetic evidence for three unique components in primary cell-wall cellulose synthase complexes in Arabidopsis. Proc Natl Acad Sci U S A 104:15566–15571

Rayle DL, Cleland RE (1992) The acid growth theory of auxin-induced cell elongation is alive and well. Plant Physiol 99:1271–1274

Refrégier G, Pelletier S, Jaillard D, Höfte H (2004) Interaction between wall deposition and cell elongation in dark-grown hypocotyl cells in Arabidopsis. Plant Physiol 135:959–968

Routier-Kierzkowska AL, Smith RS (2013) Measuring the mechanics of morphogenesis. Curr Opin Plant Biol 16:25–32

Rubery PH, Sheldrake AR (1974) Carrier-mediated auxin transport. Planta 118:101–121

Sachs J (1878) Ueber die Anordnung der Zellen in jüngsten Pflanzentheilen. Arbeiten d Bot Inst 46–104

Saladie M, Rose JK, Cosgrove DJ, Catala C (2006) Characterization of a new xyloglucan endotransglucosylase/hydrolase (XTH) from ripening tomato fruit and implications for the diverse modes of enzymic action. Plant J 47:282–295

Sampathkumar A, Gutierrez R, Mcfarlane HE, Bringmann M, Lindeboom J, Emons AM, Samuels L, Ketelaar T, Ehrhardt DW, Persson S (2013) Patterning and lifetime of plasma membrane-localized cellulose synthase is dependent on actin organization in Arabidopsis interphase cells. Plant Physiol 162:675–688

Sampathkumar A, Krupinski P, Wightman R, Milani P, Berquand A, Boudaoud A, Hamant O, Jonsson H, Meyerowitz EM (2014a) Subcellular and supracellular mechanical stress prescribes cytoskeleton behavior in Arabidopsis cotyledon pavement cells. elife 3:e01967

Sampathkumar A, Yan A, Krupinski P, Meyerowitz EM (2014b) Physical forces regulate plant development and morphogenesis. Curr Biol 24:R475–R483

Sampathkumar A, Peaucelle A, Fujita M, Schuster C, Persson S, Wasteneys GO, Meyerowitz EM (2019) Primary wall cellulose synthase regulates shoot apical meristem mechanics and growth. Development 146(10). https://doi.org/10.1242/dev.179036

Sassi M, Ali O, Boudon F, Cloarec G, Abad U, Cellier C, Chen X, Gilles B, Milani P, Friml J (2014) An auxin-mediated shift toward growth isotropy promotes organ formation at the shoot meristem in Arabidopsis. Curr Biol 24:2335–2342

Scheller HV, Ulvskov P (2010) Hemicelluloses. Annu Rev Plant Biol 61:263–289

Smith LG (2001) Plant cell division: building walls in the right places. Nat Rev Mol Cell Biol 2:33–39

Sugimoto K, Williamson RE, Wasteneys GO (2001) Wall architecture in the cellulose-deficientrsw1 mutant of Arabidopsis thaliana: microfibrils but not microtubules lose their transverse alignment before microfibrils become unrecognizable in the mitotic and elongation zones of roots. Protoplasma 215:172–183

Taylor-Teeples M, Lanctot A, Nemhauser JL (2016) As above, so below: auxin's role in lateral organ development. Dev Biol 419:156–164

Thomas LH, Forsyth VT, Sturcova A, Kennedy CJ, May RP, Altaner CM, Apperley DC, Wess TJ, Jarvis MC (2013) Structure of cellulose microfibrils in primary cell walls from collenchyma. Plant Physiol 161:465–476

Uyttewaal M, Burian A, Alim K, Landrein B, Borowska-Wykret D, Dedieu A, Peaucelle A, Ludynia M, Traas J, Boudaoud A, Kwiatkowska D, Hamant O (2012) Mechanical stress acts via katanin to amplify differences in growth rate between adjacent cells in Arabidopsis. Cell 149:439–451

Vanstraelen M, Van Damme D, De Rycke R, Mylle E, Inze D, Geelen D (2006) Cell cycle-dependent targeting of a kinesin at the plasma membrane demarcates the division site in plant cells. Curr Biol 16:308–314

Wang T, Park YB, Caporini MA, Rosay M, Zhong L, Cosgrove DJ, Hong M (2013) Sensitivity-enhanced solid-state NMR detection of expansin's target in plant cell walls. Proc Natl Acad Sci U S A 110:16444–16449

Wu S-Z, Bezanilla M (2018) Actin and microtubule cross talk mediates persistent polarized growth. J Cell Biol 217:3531–3544

Xiao C, Zhang T, Zheng Y, Cosgrove DJ, Anderson CT (2016) Xyloglucan deficiency disrupts microtubule stability and cellulose biosynthesis in Arabidopsis, altering cell growth and morphogenesis. Plant Physiol 170:234–249

Yanagisawa M, Desyatova AS, Belteton SA, Mallery EL, Turner JA, Szymanski DB (2015) Patterning mechanisms of cytoskeletal and cell wall systems during leaf trichome morphogenesis. Nat Plants 1:15014

Chapter 3
Chloroplast Actin Filaments Involved in Chloroplast Photorelocation Movements

Masamitsu Wada and Sam-Geun Kong

Abstract Plants have evolved sophisticated mechanisms to survive in various environmental changes. Chloroplast movement is an essential response to optimize photosynthesis and to avoid photodamage under fluctuating light conditions. Chloroplasts accumulate at periclinal walls to maximize light absorption under weak light while they move to anticlinal walls to minimize light exposure under strong light. The light strength is monitored by blue light receptor phototropins in general. In *Arabidopsis thaliana*, both phototropin1 (phot1) and phototropin2 (phot2) are involved in accumulation response, but phot2 is specifically involved in avoidance response. Such appropriate photorelocation movements of chloroplasts are mediated by a structure made of short actin filaments specialized for chloroplast movement. The short actin filaments are dynamically reorganized on the leading edges of moving chloroplasts, so that named chloroplast actin (cp-actin) filaments. In this chapter, we summarize recent knowledge about cp-actin filaments and next challenges to elucidate the underlying mechanisms.

3.1 Introduction

Chloroplasts show intracellular relocation movement depending on light conditions (Banaś et al. 2012; Wada 2013, 2016; Wada et al. 2003; Wada and Kong 2018). Under low light conditions chloroplasts move to the area with most light to raise photosynthetic efficiency (accumulation response or accumulation movement, or low light response), but under strong light, they move away from the light-irradiated area to avoid light-induced damage (avoidance response or avoidance movement, or high light response). These phenomena have been recognized since the nineteenth century

M. Wada (✉)
Department of Biological Sciences, Graduate School of Science, Tokyo Metropolitan University, Tokyo, Japan

S.-G. Kong
Department of Biological Sciences, College of Natural Sciences, Kongju National University, Chungnam, South Korea
e-mail: kong@kongju.ac.kr

© Springer Nature Switzerland AG 2019
V. P. Sahi, F. Baluška (eds.), *The Cytoskeleton*, Plant Cell Monographs 24,
https://doi.org/10.1007/978-3-030-33528-1_3

(Senn 1908). Studies to analyze the mechanisms more precisely started in the 1950s, mainly using a duckweed (*Lemna*) with many small chloroplasts per cell (Zurzycki 1955), and a green alga (*Mougeotia*) which has a single ribbon-shaped chloroplast per cell (Haupt 1956). The involvement of cytoskeletal filaments (actin filaments and/or microtubules) has been discussed repeatedly (Anielska-Mazur et al. 2009; Kandasamy and Meagher 1999; Krzeszowiec and Gabryś 2007; Krzeszowiec et al. 2007; Kumatani et al. 2006; Sakai and Takagi 2005; Sato et al. 2001; Takagi 2000, 2003; Takagi et al. 2009; Takamatsu and Takagi 2011), but we had to wait until very recently to obtain reliable evidence for this, in the form of photographs of chloroplast actin filaments (abbreviated as cp-actin filaments hereafter) highly correlated with chloroplast movement (Kadota et al. 2009; Kong et al. 2013; Wada and Kong 2018).

3.1.1   Chloroplast Actin Filaments

In transgenic plants of *Arabidopsis thaliana* expressing green fluorescent protein (GFP)-mouse talin that binds to filamentous actin, more short actin filaments (cp-actin filaments) were observed at the front than at the rear of moving chloroplasts when avoidance and accumulation responses were induced by high- and low-intensity lights, respectively (Fig. 3.1). The cp-actin filaments were located between the chloroplasts and the plasma membrane (Kadota et al. 2009). This biased distribution of cp-actin filaments between the front and rear of moving chloroplasts was clearly found even in mutants of photoreceptors for chloroplast movement (*phot1* for accumulation and *phot2* mainly for avoidance), but not in *phot1phot2* double mutants, in which chloroplast movement is absent (Kadota et al. 2009). Cp-actin filaments always exist at the front of a moving chloroplast, even if it rotates without changing direction (i.e., a new part of the chloroplast periphery gradually becomes the front) during the movement or if the direction of movement is changed (Kong et al. 2013), suggesting that the cp-actin filaments are controlling the direction of movement. However, since the timings of the start of chloroplast movement and that of redistribution of cp-actin filaments are almost the same, and since it is difficult to determine when chloroplasts start moving (because chloroplast movement is very slow), it is hard to say whether the redistribution of cp-actin filaments is the cause or the result of chloroplast movement. Yamada and his colleagues found that chloroplast avoidance movement was suppressed by adding 25 mM 2,3-butanedione monoxime (BDM) or N-ethylmaleimide (NEM), both potent inhibitors of myosin ATPase. After treatment, a blue light microbeam (377 μmol m^{-2} s^{-1}) was applied and distribution of cp-actin was observed. Even though chloroplast movement did not occur, biased cp-actin distribution was established, but this disappeared when the microbeam was turned off. This result indicates that movement of the chloroplast is not the cause of the biased distribution of cp-actin filaments (Yamada et al. 2011).

Cp-actin filaments disappear when irradiated with strong light by severing and then depolymerization (Kadota et al. 2009; Kong et al. 2013). This is very clear during the avoidance response induced by half-side illumination of chloroplasts with strong light.

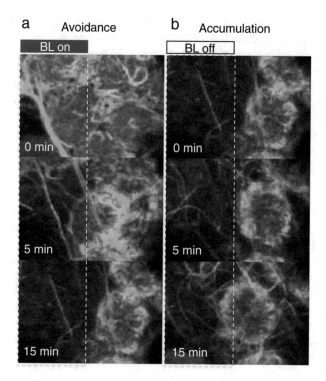

Fig. 3.1 Chloroplast actin filaments during avoidance or accumulation responses in *Arabidopsis thaliana*. (**a**) When part of a cell was irradiated with strong light ($>10\ \mu mol\ m^{-2}\ s^{-1}$; the area shown under the blue bar), chloroplasts showed the avoidance response by moving out of the irradiated area into the nonirradiated area. (**b**) When the strong light was switched off, chloroplasts showed the accumulation response by moving back into the formerly irradiated area. Green: GFP-talin showing actin filaments, red: chlorophyll fluorescence showing chloroplast position. Cp-actin filaments could be seen at the front of the moving chloroplasts during the avoidance response and around the whole periphery during the accumulation response. Modified from Fig. 4 in Kong et al. (2013) with permission

Cp-actin filaments disappear at the irradiated side and appear at the nonirradiated side. The difference in the quantity of cp-actin filaments between the front and rear determines the velocity of the avoidance movement; the bigger the difference, the higher the speed of movement (Kadota et al. 2009). The cp-actin filaments fluctuate and show a rapid turnover, so that precise analysis of these filaments using time-lapse recording of images every 30 s (Kong et al. 2013), or even at 3 or 5 s intervals (Ichikawa et al. 2011), is impossible.

The timings of cp-actin reorganization during the avoidance response induced by strong blue light irradiation have been analyzed with and without background red light (Ichikawa et al. 2011), because it has been reported that *phyA* or *phyB* mutants exhibited an enhanced avoidance response probably by modulating the transition between the low and high light-dependent chloroplast responses (DeBlasio et al. 2003). Under moderate background red light ($89\ \mu mol\ m^{-2}\ s^{-1}$), when whole cells were continuously irradiated with strong blue light, cp-actin filaments disappeared at about 30 s but reappeared, with a biased distribution, at about 70 s. The timing of the reappearance of the cp-actin filaments (with biased distribution) was delayed under strong red light or without red light, resulting in delayed chloroplast movement (Ichikawa et al. 2011). Similarly, in *phot1* mutants, the disappearance and reappearance of cp-actin filaments with biased distribution were enhanced, resulting in the avoidance response beginning sooner than in wild-type plants. This is probably because phot1 inhibits reorganization of cp-actin filaments (Ichikawa et al. 2011).

Partial irradiation of mesophyll cells by strong light induces the avoidance response, and if the light is switched off chloroplasts move back toward their original position by the accumulation response. During the return movement, cp-actin filaments appear at what is now the front (which had been the rear immediately beforehand, where cp-actin filaments were not observed), but the cp-actin filaments existing at what is now the rear do not entirely disappear. This results in only a moderate difference in the quantity of cp-actin filaments between the new front and the rear, meaning that the velocity of movement is rather slow (Kong et al. 2013) (see Fig. 3.1). In fact, while the velocity of avoidance movements is dependent on the fluence rate, the velocity of the accumulation response is almost constant irrespective of the fluence rate (Higa et al. 2017; Ichikawa et al. 2011; Kagawa and Wada 1996, 2004; Tsuboi and Wada 2011).

Cp-actin filaments are not found in the cells of some mutant plants defective in certain genes related to chloroplast movement, such as *chup1* (Kadota et al. 2009; Kong et al. 2013) and *kac1* (Suetsugu et al. 2010b).

3.1.2 Model of Motive Force Generation

The existence of cp-actin filaments and their involvement in chloroplast movement are undisputed, but how the cp-actin filaments raise the motive force for the movement is an open question. The common feature of motive force generation in the movements of plant organelles is the actomyosin system, the relationship between actin filaments and myosin molecules (Duan and Tominaga 2018; Sparkes 2011). The association of long cytoplasmic actin filaments with chloroplasts (Anielska-Mazur et al. 2009; Kadota and Wada 1992a; Kandasamy and Meagher 1999; Whippo et al. 2011) and myosin localization on plastids (Krzeszowiec and Gabryś 2007; Sattarzadeh et al. 2009) have been reported. Moreover, potent myosin ATPase inhibitors, such as NEM and BDM, affect chloroplast movement (Kadota and Wada 1992b; Kong et al. 2013; Paves and Truve 2007; Yamada et al. 2011). However, the involvement of myosins is controversial. It is possible that NEM and BDM inhibit proteins other than myosin (Yamada et al. 2011). In *Arabidopsis* myosin XI double, triple, and quadruple knockout lines, normal accumulation and avoidance responses of chloroplasts have been observed, although other organelle movements are blocked (Suetsugu et al. 2010a). The involvement of myosins XI and VIII other than those knocked out is still a possibility. In *Physcomitrella patens*, reorganization of cp-actin filaments after light irradiation to induce accumulation and avoidance responses can be seen even after treatment with BDM or the microtubule inhibitor Oryzalin (Yamashita et al. 2011).

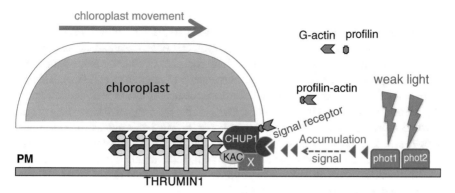

Fig. 3.2 A model of chloroplast movement based on current knowledge assuming motive force generation by cp-actin polymerization in the peripheral region of chloroplast. Cp-actin filaments polymerized by a CHUP1 complex are anchored to the plasma membrane with THRUMIN1, so that newly polymerized actin filaments between CHUP1 and the former cp-actin filaments push the CHUP1 complex at the front of the chloroplast forward, resulting in chloroplast movement. Cp-actin polymerization continues as long as signals released from phot1 and phot2 reach the CHUP1 complex

Polymerization of cp-actin filaments might be the most likely mechanism explaining the generation of motive force for chloroplast movement, in the manner of the "actin comet tails," at the rear of bacteria within host cells, that generate force by actin filament polymerization by the actin-related protein 2/3 (Arp2/3) complex (Goley and Welch 2006). However, Arp2/3 is not the cp-actin polymerization factor here, because Arp2/3-deficient mutants of *Arabidopsis* exhibit normal chloroplast movements (Kadota et al. 2009). If polymerization of cp-actin filaments generates the motive force for chloroplast movement, the polymerized actin filaments would need to be anchored to the plasma membrane. Otherwise, the filaments would drift freely and could not function to generate a force (Fig. 3.2). Actin bundling factor THRUMIN1, a plasma membrane protein (Whippo et al. 2011), might be the candidate. A transgenic plant expressing both THRUMIN1-RFP and GFP-talin has been shown to exhibit similar fluorescence distribution patterns for both (Kong et al. 2013), suggesting chloroplast anchorage to the plasma membrane in collaboration with THRUMIN1 and cp-actin filaments.

3.1.3   Physcomitrella *Is an Exceptional Case*

Chloroplast movement during both accumulation and avoidance responses can be induced by blue as well as red light in protonemata of *P. patens*, although the red light effect can be seen only when the protonemata are cultured under red light conditions. If the red light-grown cells are further cultured under white light for 2 days, the red light response is lost. The red light effects are also nullified by a simultaneous background irradiation of far-red light, indicating phytochrome involvement as the photoreceptor (Kadota et al. 2000). The blue light effects are mediated by two groups of phototropins (photA and photB groups) (Kasahara et al. 2004). Interestingly, the red light-induced chloroplast movement is also significantly reduced in *photA2photB1photB2* triple disruptants (Kasahara et al. 2004). This can be explained by the fact that a phytochrome associates physically with phototropin at the plasma membrane (Jaedicke et al. 2012). The light signal received by phytochrome must be transferred to phototropins as a signal transducer for chloroplast movement. This physical interaction and the signal transfer between the two photoreceptors are also found as a good example of the convergent evolution in chimeric proteins neochromes consisting of a phytochrome N-terminal chromophore-binding domain and a full-length of phototropin found in the fern *Adiantum capillus-veneris* (Kawai et al. 2003; Nozue et al. 1998) and the green alga *Mougeotia scalaris* (Suetsugu et al. 2005). Neochromes have originated twice independently: once in zygnematalean algae (such as *Mougeotia*) and once in hornworts (Li and Mathews 2016; Suetsugu et al. 2005). The hornwort neochrome was transferred horizontally to ferns (Li et al. 2014).

To study the mechanism of chloroplast movement in *P. patens*, red light-grown protonemata were kept in darkness for 1 day to strengthen their sensitivity to light, then cytoskeletal inhibitors (100 μM cytochalasin B or 10 μM Latrunculin B for microfilaments (i.e., actin filaments) and 10 μM Cremart or 10 μM Oryzalin for microtubules) were applied (Sato et al. 2001). It was found that both microfilaments and microtubules are simultaneously involved in the movements. In the dark without induction of chloroplast movement, microtubules mediated rapid chloroplast movement back and forth along the longitudinal cell axis and microfilaments mediated slow movement in any direction. Red light-induced movements for both accumulation and avoidance responses were completely blocked by Cremart but not by cytochalasin B, indicating a phytochrome-mediated avoidance response depending on microtubules. Blue light-induced movements were inhibited when both Cremart and cytochalasin B were simultaneously applied, indicating that both microtubules and actin filaments are involved. Interestingly, chloroplast behaviors mediated by microtubules and actin filaments are different. Actin filaments mediate straight movement in any direction even to the cell side, but microtubules induce back and forth movement along the cell axis to the final destination (Sato et al. 2001). The actin-based movement was mediated by CHUP1 (Usami et al. 2012).

3.1.4   Inconsistency of the Model

When only a part of an *Arabidopsis* leaf cell (about 10 μm in width) is irradiated with a high fluence rate of blue light using a microbeam irradiator, chloroplasts only semi-irradiated escape rapidly outside of the beam, and these show cp-actin filaments at the front of chloroplasts (Kadota et al. 2009; Kong et al. 2013). However, chloroplasts inside the beam stay in the beam-irradiated area and do not move away from the strong light at least for a while (Kagawa and Wada 2004), because the cp-actin filaments depolymerize completely (Kong et al. 2013). In nature, however, when whole leaves are illuminated with strong sunlight, all the cp-actin filaments must be depolymerized. In this case, an escape reaction might not occur, but in fact it does. Why?

Based on our experiments using *Arabidopsis* knockout lines of four myosin XI genes that are mainly expressed in leaves, myosin XI might not be involved in chloroplast movement, because normal accumulation and avoidance responses of the chloroplasts were observed, even though other organelle movements were deficient (Suetsugu et al. 2010a). However, we did not knock out all the myosin genes simultaneously, and so the possibility of myosin involvement remains. Moreover, the myosin inhibitors NEM and BDM effectively inhibit chloroplast movement (Kong et al. 2013; Paves and Truve 2007; Yamada et al. 2011). How do we align these experimental results with our model?

It is quite curious that in *kac1kac2* double mutants, where no cp-actin filaments can be observed, the avoidance response occurs when detected by leaf transmittance of red light (Suetsugu et al. 2016). KAC1 and KAC2, and CHUP1, are necessary for polymerization and/or maintenance of cp-actin filaments (Kadota et al. 2009; Kong et al. 2013; Suetsugu et al. 2010b, 2012). Are there any ways to induce chloroplast movement other than via cp-actin filaments in seed plants?

3.2 Issues to Be Clarified in the Near Future

3.2.1 Biased Distribution of cp-Actin Filaments

Chloroplasts can move in any direction, meaning that they have no innate front and back. Their polarity (head and tail) for chloroplast movement might be decided just before the start of movement; the peripheral region nearest to its destination becomes the front of the chloroplast. The speed of chloroplast movement depends on the difference in quantity of cp-actin filaments between the front and rear, and it is this biased distribution that is the most important factor (Kadota et al. 2009; Kong et al. 2013). The mechanism of establishment of the bias is different between the accumulation and avoidance responses. In an avoidance response induced by strong light, cp-actin filaments initially disappear from all parts of the chloroplast, but then a dense mass of new cp-actin filaments appears only at the front, resulting in a large difference between front and rear (Kadota et al. 2009). In an accumulation response, cp-actin filaments do not disappear; new cp-actin filaments are polymerized at the front in addition to the old ones, and this results in only a slight difference between front and rear, and rather slow speed of movement (Kong et al. 2013). In both cases, cp-actin filaments are polymerized at the front of moving chloroplasts to create a distribution bias. How is this biased distribution pattern of cp-actin filaments established? Biased activation of cp-actin polymerizing factor(s) along the chloroplast periphery must be a possible explanation.

3.2.2 Analysis of Rapidly Moving Cp-Actin Filaments

Cp-actin filaments move very rapidly. In an early study, we took time-lapse photographs every 5 min and identified cp-actin filaments (Kadota et al. 2009), but analysis of the cp-actin filament behavior was impossible. We then took time-lapse images every 30 s (Kong et al. 2013), but it was not possible to identify and match up the same filament in two serial frames for comparison.

Depolymerization of cp-actin filaments was observed by total internal reflection fluorescence microscopy, and time-lapse frames were acquired every 0.2 s (5 frames per second) (Kong et al. 2013), but even then precise analysis of the severing of cp-actin filaments was very difficult. Accordingly, it will be necessary to take time-lapse photographs with even higher temporal resolutions, and then hopefully it will be possible to analyze the behavior of individual cp-actin filaments and apply that knowledge to increase the reliability of our model.

3.2.3 Polymerization Factor(s)

In *chup1* and *kac1* mutant leaves, chloroplast movements are deficient (Oikawa et al. 2003; Oikawa et al. 2008; Suetsugu et al. 2010b), and cp-actin filaments cannot be detected (Kadota et al. 2009; Kong et al. 2013; Suetsugu et al. 2010b), suggesting the involvement of CHUP1 and KACs for cp-actin polymerization and/or maintenance. Both CHUP1 and KACs exist in the plant kingdom only above Charophytes (Suetsugu and Wada 2016), therefore if either or both are involved in cp-actin filament polymerization as an actin nucleation factor(s), plants have evolved a plant-specific, unique actin polymerization system. Precise biochemical studies are needed to prove this.

As actin nucleation factors, the Arp2/3 complex (Goley and Welch 2006), formins (Chalkia et al. 2008), and tandem-monomer-binding nucleators, including Spire (Firat-Karalar and Welch 2011), have been reported. The *Arabidopsis* mutant line of *arp2/3*, however, shows normal chloroplast movements (Kadota et al. 2009). Plant formins have diverse isoforms, with >20 isoforms in the *Arabidopsis* genome. *Arabidopsis* formins are widely expressed and regulate diverse cellular responses including cytokinesis, cell-to-cell trafficking (Diao et al. 2018), pollen cell growth, and cell expansion (Deeks et al. 2005; Ingouff et al. 2005; Lan et al. 2018; Li et al. 2010; Ye et al. 2009). The formin proteins are capable of actin assembly by incorporating G-actin monomers into the barbed end of a filament using the conserved C-terminal FH1 and FH2 domains (Blanchoin and Staiger 2010; van Gisbergen and Bezanilla 2013; Wang et al. 2012). However, there is no information available on whether or not formins are involved in chloroplast movement.

3.3 Motive Force Generation

To clarify the mechanism of motive force generation in chloroplast movement, the precise structure of cp-actin filaments at electron microscope level and individual behavior of a cp-actin filament should be studied, although both experiments would be technically challenging. If we could observe the cp-actin filaments of moving chloroplasts that had been fixed by a high-pressure rapid freezing method, the involvement (or otherwise) of an actomyosin system might be determined. Another possibility might be an in vitro assay using CHUP1 and/or KACs that might act as cp-actin polymerization factors. When in vitro polymerization of cp-actin filaments becomes a reality, cp-actin filament behavior could be analyzed using atomic force microscopy.

3.4 Concluding Remarks

We believe that the motive force generated by cp-actin filaments is the main mechanism of chloroplast movement in land plants, and we consider that myosins are not involved. However, as mentioned above, to prove this hypothesis or to otherwise fully clarify the model for chloroplast movement, persistent efforts will be required to overcome the current limitations in microscopy, biochemical analysis, and crystallography, and to develop new techniques in collaboration with specialists in those fields.

Acknowledgment The work was supported by the grants from the Japan Society for the Promotion of Science (JSPS) (No. 20227001, 23120523, 25120721, 25251033, and 16K14758) and from Ohsumi Frontier Science Foundation to M.W. and by Basic Science Research Program through the National Research Foundation of Korea (NRF) funded by the Ministry of Education (No. 2016R1D1A3B03935947) and the Next-Generation BioGreen 21 Program grant funded by the Korea government Rural Development Administration (RDA) (No. PJ01366901) to S.-G. K.

References

Anielska-Mazur A, Bernaś T, Gabryś H (2009) In vivo reorganization of the actin cytoskeleton in leaves of *Nicotiana tabacum* L. transformed with plastin-GFP. Correlation with light-activated chloroplast responses. BMC Plant Biol 9:64

Banaś AK, Aggarwal C, Łabuz J, Sztatelman O, Gabryś H (2012) Blue light signalling in chloroplast movements. J Exp Bot 63:1559–1574

Blanchoin L, Staiger CJ (2010) Plant formins: diverse isoforms and unique molecular mechanism. Biochim Biophys Acta 1803:201–206

Chalkia D, Nikolaidis N, Makalowski W, Klein J, Nei M (2008) Origins and evolution of the formin multigene family that is involved in the formation of actin filaments. Mol Biol Evol 25:2717–2733

DeBlasio SL, Mullen JL, Luesse DR, Hangarter RP (2003) Phytochrome modulation of blue light-induced chloroplast movements in *Arabidopsis*. Plant Physiol 133:1471–1479

Deeks MJ, Cvrckova F, Machesky LM, Mikitova V, Ketelaar T, Zarsky V, Davies B, Hussey PJ (2005) *Arabidopsis* group Ie formins localize to specific cell membrane domains, interact with actin-binding proteins and cause defects in cell expansion upon aberrant expression. New Phytol 168:529–540

Diao M, Ren S, Wang Q, Qian L, Shen J, Liu Y, Huang S (2018) Arabidopsis formin 2 regulates cell-to-cell trafficking by capping and stabilizing actin filaments at plasmodesmata. elife 7: e36316

Duan Z, Tominaga M (2018) Actin-myosin XI: an intracellular control network in plants. Biochem Biophys Res Commun 506:403–408

Firat-Karalar EN, Welch MD (2011) New mechanisms and functions of actin nucleation. Curr Opin Cell Biol 23:4–13

Goley ED, Welch MD (2006) The ARP2/3 complex: an actin nucleator comes of age. Nat Rev Mol Cell Biol 7:713–726

Haupt W (1956) Chloroplastenbewegung. Z Bot 44:455–462

Higa T, Hasegawa S, Hayasaki Y, Kodama Y, Wada M (2017) Temperature-dependent signal transmission in chloroplast accumulation response. J Plant Res 130:779–789

Ichikawa S, Yamada N, Suetsugu N, Wada M, Kadota A (2011) Red light, phot1 and JAC1 modulate phot2-dependent reorganization of chloroplast actin filaments and chloroplast avoidance movement. Plant Cell Physiol 52:1422–1432

Ingouff M, Fitz Gerald JN, Guerin C, Robert H, Sorensen MB, Van Damme D, Geelen D, Blanchoin L, Berger F (2005) Plant formin AtFH5 is an evolutionarily conserved actin nucleator involved in cytokinesis. Nat Cell Biol 7:374–380

Jaedicke K, Lichtenthaler AL, Meyberg R, Zeidler M, Hughes J (2012) A phytochrome-phototropin light signaling complex at the plasma membrane. Proc Natl Acad Sci U S A 109:12231–12236

Kadota A, Wada M (1992a) Photoinduction of formation of circular structures by microfilaments on chloroplasts during intracellular orientation in protonemal cells of the fern *Adiantum capillus-veneris*. Protoplasma 167:97–107

Kadota A, Wada M (1992b) Photoorientation of chloroplasts in protonemal cells of the fern *Adiantum* as analyzed by use of a video-tracking system. Bot Mag Tokyo 105:265–279

Kadota A, Sato Y, Wada M (2000) Intracellular chloroplast photorelocation in the moss *Physcomitrella patens* is mediated by phytochrome as well as by a blue-light receptor. Planta 210:932–937

Kadota A, Yamada N, Suetsugu N, Hirose M, Saito C, Shoda K, Ichikawa S, Kagawa T, Nakano A, Wada M (2009) Short actin-based mechanism for light-directed chloroplast movement in *Arabidopsis*. Proc Natl Acad Sci U S A 106:13106–13111

Kagawa T, Wada M (1996) Phytochrome- and blue-light-absorbing pigment-mediated directional movement of chloroplasts in dark-adapted prothallial cells of fern *Adiantum* as analyzed by microbeam irradiation. Planta 198:488–493

Kagawa T, Wada M (2004) Velocity of chloroplast avoidance movement is fluence rate dependent. Photochem Photobiol Sci 3:592–595

Kandasamy MK, Meagher RB (1999) Actin-organelle interaction: association with chloroplast in Arabidopsis leaf mesophyll cells. Cell Motil Cytoskeleton 44:110–118

Kasahara M, Kagawa T, Sato Y, Kiyosue T, Wada M (2004) Phototropins mediate blue and red light-induced chloroplast movements in *Physcomitrella patens*. Plant Physiol 135:1388–1397

Kawai H, Kanegae T, Christensen S, Kiyosue T, Sato Y, Imaizumi T, Kadota A, Wada M (2003) Responses of ferns to red light are mediated by an unconventional photoreceptor. Nature 421:287–290

Kong S-G, Arai Y, Suetsugu N, Yanagida T, Wada M (2013) Rapid severing and motility of chloroplast-actin filaments are required for the chloroplast avoidance response in *Arabidopsis*. Plant Cell 25:572–590

Krzeszowiec W, Gabryś H (2007) Phototropin mediated relocation of myosins in *Arabidopsis thaliana*. Plant Signal Behav 2:333–336

Krzeszowiec W, Rajwa B, Dobrucki J, Gabryś H (2007) Actin cytoskeleton in *Arabidopsis thaliana* under blue and red light. Biol Cell 99:251–260

Kumatani T, Sakurai-Ozato N, Miyawaki N, Yokota E, Shimmen T, Terashima I, Takagi S (2006) Possible association of actin filaments with chloroplasts of spinach mesophyll cells *in vivo* and *in vitro*. Protoplasma 229:45–52

Lan Y, Liu X, Fu Y, Huang S (2018) Arabidopsis class I formins control membrane-originated actin polymerization at pollen tube tips. PLoS Genet 14:e1007789

Li FW, Mathews S (2016) Evolutionary aspects of plant photoreceptors. J Plant Res 129:115–122

Li Y, Shen Y, Cai C, Zhong C, Zhu L, Yuan M, Ren H (2010) The type II *Arabidopsis* formin14 interacts with microtubules and microfilaments to regulate cell division. Plant Cell 22:2710–2726

Li FW, Villarreal JC, Kelly S, Rothfels CJ, Melkonian M, Frangedakis E, Ruhsam M, Sigel EM, Der JP, Pittermann J, Burge DO, Pokorny L, Larsson A, Chen T, Weststrand S, Thomas P, Carpenter E, Zhang Y, Tian Z, Chen L, Yan Z, Zhu Y, Sun X, Wang J, Stevenson DW, Crandall-Stotler BJ, Shaw AJ, Deyholos MK, Soltis DE, Graham SW, Windham MD, Langdale JA, Wong GK, Mathews S, Pryer KM (2014) Horizontal transfer of an adaptive chimeric photoreceptor from bryophytes to ferns. Proc Natl Acad Sci U S A 111:6672–6677

Nozue K, Kanegae T, Imaizumi T, Fukuda S, Okamoto H, Yeh KC, Lagarias JC, Wada M (1998) A phytochrome from the fern *Adiantum* with features of the putative photoreceptor *NPH1*. Proc Natl Acad Sci U S A 95:15826–15830

Oikawa K, Kasahara M, Kiyosue T, Kagawa T, Suetsugu N, Takahashi F, Kanegae T, Niwa Y, Kadota A, Wada M (2003) Chloroplast unusual positioning1 is essential for proper chloroplast positioning. Plant Cell 15:2805–2815

Oikawa K, Yamasato A, Kong S-G, Kasahara M, Nakai M, Takahashi F, Ogura Y, Kagawa T, Wada M (2008) Chloroplast outer envelope protein CHUP1 is essential for chloroplast anchorage to the plasma membrane and chloroplast movement. Plant Physiol 148:829–842

Paves H, Truve E (2007) Myosin inhibitors block accumulation movement of chloroplasts in *Arabidopsis thaliana* leaf cells. Protoplasma 230:165–169

Sakai Y, Takagi S (2005) Reorganized actin filaments anchor chloroplasts along the anticlinal walls of Vallisneria epidermal cells under high-intensity blue light. Planta 221:823–830

Sato Y, Wada M, Kadota A (2001) Choice of tracks, microtubules and/or actin filaments for chloroplast photo-movement is differentially controlled by phytochrome and a blue light receptor. J Cell Sci 114:269–279

Sattarzadeh A, Krahmer J, Germain AD, Hanson MR (2009) A myosin XI tail domain homologous to the yeast myosin vacuole-binding domain interacts with plastids and stromules in Nicotiana benthamiana. Mol Plant 2(6):1351–1358

Senn G (1908) Die Gestalts- und Lageveränderung der Pflanzen-Chromatophoren. Wilhelm Engelmann, Leipzig

Sparkes I (2011) Recent advances in understanding plant myosin function: life in the fast lane. Mol Plant 4:805–812

Suetsugu N, Wada M (2016) Evolution of the cp-actin-based motility system of chloroplasts in green plants. Front Plant Sci 7:561

Suetsugu N, Mittmann F, Wagner G, Hughes J, Wada M (2005) A chimeric photoreceptor gene, NEOCHROME, has arisen twice during plant evolution. Proc Natl Acad Sci U S A 102:13705–13709

Suetsugu N, Dolja VV, Wada M (2010a) Why have chloroplasts developed a unique motility system? Plant Signal Behav 5:1190–1196

Suetsugu N, Yamada N, Kagawa T, Yonekura H, Uyeda TQP, Kadota A, Wada M (2010b) Two kinesin-like proteins mediate actin-based chloroplast movement in *Arabidopsis thaliana*. Proc Natl Acad Sci U S A 107:8860–8865

Suetsugu N, Sato Y, Tsuboi H, Kasahara M, Imaizumi T, Kagawa T, Hiwatashi Y, Hasebe M, Wada M (2012) The KAC family of kinesin-like proteins is essential for the association of chloroplasts with the plasma membrane in land plants. Plant Cell Physiol 53:1854–1865

Suetsugu N, Higa T, Gotoh E, Wada M (2016) Light-induced movements of chloroplasts and nuclei are regulated in both cp-actin-filament-dependent and -independent manners in *Arabidopsis thaliana*. PLoS One 11:e0157429

Takagi S (2000) Roles for actin filaments in chloroplast motility and anchoring. In: Staiger CJ, Baluška F, Volkmann D, Barlow PW (eds) Actin: a dynamic framework for multiple plant cell functions. Kluwer Academic, Dordrecht, The Netherlands, pp 203–212

Takagi S (2003) Actin-based photo-orientation movement of chloroplasts in plant cells. J Exp Biol 206:1963–1969

Takagi S, Takamatsu H, Sakurai-Ozato N (2009) Chloroplast anchoring: its implications for the regulation of intracellular chloroplast distribution. J Exp Bot 60:3301–3310

Takamatsu H, Takagi S (2011) Actin-dependent chloroplast anchoring is regulated by Ca^{2+}-calmodulin in spinach mesophyll cells. Plant Cell Physiol 52:1973–1982

Tsuboi H, Wada M (2011) Chloroplasts can move in any direction to avoid strong light. J Plant Res 124:201–210

Usami H, Maeda T, Fujii Y, Oikawa K, Takahashi F, Kagawa T, Wada M, Kasahara M (2012) CHUP1 mediates actin-based light-induced chloroplast avoidance movement in the moss *Physcomitrella patens*. Planta 236:1889–1897

van Gisbergen PA, Bezanilla M (2013) Plant formins: membrane anchors for actin polymerization. Trends Cell Biol 23:227–233

Wada M (2013) Chloroplast movement. Plant Sci 210:177–182

Wada M (2016) Chloroplast and nuclear photorelocation movements. Proc Jpn Acad Ser B Phys Biol Sci 92:387–411

Wada M, Kagawa T, Sato Y (2003) Chloroplast movement. Annu Rev Plant Biol 54:455–468

Wada M, Kong S-G (2018) Actin-mediated movement of chloroplasts. J Cell Sci 131:jcs210310

Wang J, Xue X, Ren H (2012) New insights into the role of plant formins: regulating the organization of the actin and microtubule cytoskeleton. Protoplasma 249(Suppl 2):S101–S107

Whippo CW, Khurana P, Davis PA, DeBlasio SL, DeSloover D, Staiger CJ, Hangarter RP (2011) THRUMIN1 is a light-regulated actin-bundling protein involved in chloroplast motility. Curr Biol 21:59–64

Yamada N, Suetsugu N, Wada M, Kadota A (2011) Phototropin-dependent biased relocalization of cp-actin filaments can be induced even when chloroplast movement is inhibited. Plant Signal Behav 6:1651–1653

Yamashita H, Sato Y, Kanegae T, Kagawa T, Wada M, Kadota A (2011) Chloroplast actin filaments organize meshwork on the photorelocated chloroplasts in the moss *Physcomitrella patens*. Planta 233:357–368

Ye J, Zheng Y, Yan A, Chen N, Wang Z, Huang S, Yang Z (2009) *Arabidopsis* formin3 directs the formation of actin cables and polarized growth in pollen tubes. Plant Cell 21:3868–3884

Zurzycki J (1955) Chloroplast arrangement as a factor in photosynthesis. Acta Soc Bot Pol 24:27–63

Chapter 4
Diversity of Plant Actin–Myosin Systems

Takeshi Haraguchi, Zhongrui Duan, Masanori Tamanaha, Kohji Ito, and Motoki Tominaga

Abstract Interactions between the actin cytoskeleton and myosin motor proteins are crucial for force generation, intracellular transport, and morphogenesis in eukaryotic cells. In plant cells, the rapid intracellular transport system—cytoplasmic streaming—is generated by the interaction between actin and the plant-specific myosin XI. Genomic analyses have revealed numerous actin and myosin genes (paralogues) in angiosperms, suggesting that the plant actin–myosin XI system is more complex than expected. Recent molecular biological and biochemical approaches have revealed the functional diversity of actins and myosins in vascular plants. Actin isoforms show various biochemical properties in vitro and form distinct filamentous structures in cells. Myosin XIs exhibit various enzymatic properties and velocities, and their classification based on velocities crudely correlates with their expression pattern in tissues. Myosin XI isoform numbers increase with the evolution of plants from algae to angiosperms, suggesting that diversity of the actin–myosin system is essential for higher plant systems, such as development, morphogenesis, fertilisation, and environmental response. In this review, we summarise recent advances in research into the plant actin–myosin system and discuss the diversity entwined with plant evolution, and then propose a new model for intracellular transport regulated by multiple actin–myosin isoforms.

T. Haraguchi · M. Tamanaha · K. Ito
Department of Biology, Graduate School of Science, Chiba University, Chiba, Japan

Z. Duan
Faculty of Education and Integrated Arts and Sciences, Waseda University, Tokyo, Japan

M. Tominaga (✉)
Faculty of Education and Integrated Arts and Sciences, Waseda University, Tokyo, Japan

Department of Integrative Bioscience and Biomedical Engineering, Graduate School of
Advanced Science and Engineering, Waseda University, Tokyo, Japan
e-mail: motominaga@waseda.jp

© Springer Nature Switzerland AG 2019
V. P. Sahi, F. Baluška (eds.), *The Cytoskeleton*, Plant Cell Monographs 24,
https://doi.org/10.1007/978-3-030-33528-1_4

4.1 Introduction

In eukaryotes, the actin–myosin system plays an important role in various biological functions, such as muscle contraction, cell division, cell motility, auditory sensing, and endocytosis. In plants, the actin–myosin system is involved in cytoplasmic streaming that is generated by organelle associated plant-specific myosin XIs moving along actin fibres in the cells. Cytoplasmic streaming has been thought to be a relatively simple system to promote the mixing of cytoplasm and cellular substances (Shimmen and Yokota 2004; Yamamoto et al. 1999). Recently, genomic analyses have revealed numerous actin and myosin isoforms in vascular plants.

Actin is present in all eukaryotic cells and plays important roles in various cellular functions. Typically, land plants possess several actin isoforms (Gunning et al. 2015; Nishiyama et al. 2018; Slajcherova et al. 2012) (Fig. 4.1). Recent data have revealed that the 8 actin isoforms in Arabidopsis show distinct biochemical properties and form distinct filamentous structures in vivo (Kijima et al. 2016, 2018).

			The number of actins	The number of myosins	
				Class VIII	Class XI
Streptophyta	Eudicotyledons	*Arabidopsis thaliana*	8[a] (5[e])	4[d] (2[f])	13[d] (7[g])
		Glycine max	17[a]	7[d]	18[d]
		Populus trichocarpa	8-9[b]	4[d]	10[d]
	Monocotyledons	*Musa acuminate*	12[a]	3[d]	11[d]
		Oryza sativa	8-10[b]	2[d]	11[d]
		Zea mays	21[b]	3[d]	11[d]
	Gymnosperms	*Pinus taeda*	10[b]	-	-
	Lycopodiophyta	*Selaginella moellendorffii*	2-3[b]	2[d]	2[d]
	Bryophyta	*Physcomitrella patens*	10[b]	5[d]	2[d]
	Charophyta	*Chara braunii*	16[c]	2[c]	4[c]
Chlorophyta		*Chlamydomonas reinhardtii*	1[a]	1[d]	2[d]

Fig. 4.1 The number of actin, myosin VIII, and XI isoforms in plants. The number of actins increased from *Chlamydomonas reinhardtii* (Chlorophyta) to *Chara braunii* (Streptophyta). Although the number of myosin VIII isoforms did not markedly increase among plants, whereas the number of myosin XI isoforms remarkably increased in angiosperms. Numbers in parentheses represent numbers of isoforms expressed in the pollen of Arabidopsis. [a]Gunning et al. (2015); [b]Slajcherova et al. (2012); [c]Nishiyama et al. (2018); [d]Muhlhausen and Kollmar (2013); [e]Meagher et al. (1999); [f]Peremyslov et al. (2010) and our unpublished data; [g]Haraguchi et al. (2018)

Myosin is a motor protein capable of producing motive force along actin filaments using the energy released through ATP hydrolysis. In all eukaryotes, at least 79 myosin classes have been identified (Kollmar and Muhlhausen 2017). Among these, plants show only 2 classes of plant-specific myosin (classes VIII and XI). Arabidopsis has 4 myosin VIII genes (VIII-A, -B, ATM1, and ATM2) and 13 myosin XI genes (XI-1, -2, -A, -B, -C, -D, -E, -F, -G, -H, -I, -J, and -K) (Reddy 2001). Myosin VIII and XI have similar domain compositions including a motor domain (MD), neck domain with isoleucine–glutamine (IQ) motifs, coiled-coil region, and globular tail domain (GTD). However, the molecular size of myosin VIII is expected to be smaller than that of myosin XI because of the shorter lever arm (comprising 3–4 IQ motifs) and shorter predicted coiled-coil region (Tominaga and Ito 2015; Tominaga and Nakano 2012). Physiological functions of myosin VIII and XI are distinct. Several studies have indicated that myosin VIII is involved in new cell wall formation, intercellular transport through plasmodesmata and endocytosis (Avisar et al. 2008; Baluska et al. 2001; Golomb et al. 2008; Haraguchi et al. 2014; Reichelt et al. 1999; Sattarzadeh et al. 2008). Myosin XI is involved in cytoplasmic streaming and organelle movements (Peremyslov et al. 2008, 2010; Prokhnevsky et al. 2008; Tominaga et al. 2013; Ueda et al. 2010), plant development (Ojangu et al. 2012; Peremyslov et al. 2010; Prokhnevsky et al. 2008; Tominaga et al. 2013), movement and shape of the nucleus (Tamura et al. 2013), plant posture and straightening (Okamoto et al. 2015), pollen tube growth (Madison et al. 2015), and resistance against fungal pathogens (Yang et al. 2014).

In contrast to animal myosins (e.g. skeletal muscle myosin II), mechanical and biochemical functions of plant myosin XI have not been elucidated to date due to the difficulty of purifying sufficient amounts of plant myosins from plant cells, in which more than 90% of the cell volume is occupied by vacuoles containing high levels of proteases. In 1994, active myosin XI was successfully purified from *Lilium longiflorum* pollen tubes (Yokota and Shimmen 1994) and from *Chara corallina* internodal cells (Yamamoto et al. 1994) using appropriate approaches to inhibit or eliminate proteases. Expression and purification of the recombinant myosin XI MD (*Chara corallina* myosin XI) was achieved using an insect cell expression system (Ito et al. 2003). Notable, the molecular morphology and movement of plant myosin XI (tobacco 175-kDa myosin XI) has been revealed at a single molecular level using electron microscopy and optical trap nanometry (Tominaga et al. 2003). So far, 17 myosin sequences have been identified in the Arabidopsis genome, which are categorised into two groups—class VIII and class XI (Reddy 2001). In general, myosins in the same class show similar enzymatic properties (El-Mezgueldi and Bagshaw 2008). Unexpectedly, a recent biochemical study has revealed variable enzymatic properties of different Arabidopsis myosin XIs (Haraguchi et al. 2018). In addition, the affinity between plant myosin and actin isoforms has been shown to differ (Rula et al. 2018). In this chapter, we review the diversity of the plant actin–myosin XI system from the viewpoint of plant evolution.

	Reproductive	Vegetative		Root
		Shoot	Leaf	
Ubiquitous				
XI-1[a]	+ Style		++ Cotyledon ++ 1st leaf +++ Stipule	+ Meristematic zone + Elongation zone
XI-2[a]	+++ Style ++ Anther +++ Pollen	+++ Shoot Meristem	+++ Cotyledon ++ 1st leaf +++ Stipule	+++ Root cap +++ Stele
XI-B[a]	+++ Ovule +++ Pollen	++ Shoot Meristem + Nod	++ 1st leaf +++ Stipule +++ Trichome	+++ Lateral root ++ Root cap ++ Stele (elongating zone)
XI-K[a]	+++ Petal +++ Style +++ Stigma +++ Ovule +++ Anther	+++ Shoot Meristem	+++ 1st leaf +++ Stipule +++ Trichome	+++ Root tip +++ Root hair
ACT7[b]	+/- Mature pollen and pollen tube	+++ Hypocotyl ++ Vascular cylinder +++ Floral or vegetative meristem	+ Young and/or mature leaves	+++ Root tip + Root cortical and epidermal tissue
ACT2[b]	+/- Mature pollen and pollen tube	+/- Hypocotyl +++ Vascular cylinder +++ Floral or vegetative meristem	+++ Young and/or mature leaves	+++ Root tip +++ Root cortical and epidermal tissue
ACT8[b]	+/- Mature pollen and pollen tube	++ Vascular cylinder ++ Floral or vegetative meristem	+++ Young and/or mature leaves	++ Root tip ++ Root cortical and epidermal tissue
Pollen				
XI-A[a]	+++ Pollen			
XI-C[a]	+++ Pollen			
XI-D[a]	+++ Pollen			
XI-E[a]	+++ Pollen			
XI-J[a]	+++ Pollen			
ACT11[b]	++ Mature pollen and pollen tube +++ Developing carpel +++ Transmittal tissue +++ Endosperm +++ Young developing ovules +++ Embryo +++ Developing seed and silique	+ Hypocotyl +/- Vascular cylinder ++ Floral or vegetative meristem		+ Root tip
ACT1[b]	+++ Mature pollen and pollen tube + Developing carpel ++ Young developing ovules + Embryo	+/- Vascular cylinder ++ Floral or vegetative meristem		+ Root tip
ACT3[b]	++ Mature pollen and pollen tube + Developing carpel ++ Young developing ovules + Embryo	+/- Vascular cylinder ++ Floral or vegetative meristem		+ Root tip
ACT4[b]	++ Mature pollen and pollen tube	+/- Vascular cylinder		+ Root tip
ACT12[b]	++ Mature pollen and pollen tube	+/- Vascular cylinder		+ Root tip
Others				
XI-F[a]		+++ Stele	++ 1st leaf (Midrib) ++ 1st leaf +++ Stipule ++ Stipule	+++ Stele +++ Root cap +++ Stele
XI-G[a]	+++ Stigma ++ Ovule			
XI-H[a]	++ Style	++ Node		
XI-I[a]	+ Anther			

Fig. 4.2 Expression pattern of At actin (red) and At myosin XI isoforms (green). At actin and At myosin XI isoforms expressed in Arabidopsis were summarised. The number of + symbols indicates the approximate expression intensity. High-, Medium-, and Low-velocity At myosin XIs were highlighted by red, blue, and green, respectively. [a]Haraguchi et al. (2018); [b]Meagher et al. (1999)

4.2 Diversity of Plant Actins

The numbers of cytoplasmic actin genes are remarkably larger in plants than in metazoa, fungi, and protists (Gunning et al. 2015). Recent genetic and biochemical studies have demonstrated that plant actins are diverse in terms of numbers as well as functions. Increase in the number of actins from *Chlamydomonas reinhardtii* (Chlorophyta) to *Chara braunii* (Streptophyta) highlights the importance of actin diversification during plant evolution from unicellular to multicellular plants (Fig. 4.1). Overall, 8 actin isoforms are expressed in Arabidopsis and divided into 2 groups: those expressed in vegetative organs (AtACT2, 7 and 8) and those expressed in reproductive organs (AtACT1, 3, 4, 11, and 12) (McDowell et al. 1996; Meagher et al. 1999) (Fig. 4.2). Gene knockout analyses have revealed roles of Arabidopsis actin isoforms in cell expansion, trichome shape, root growth, and organ straightening and plant posture (Kandasamy et al. 2009; Kato et al. 2010; Lanza et al. 2012).

The amino acid sequence identities and similarities among Arabidopsis actin isoforms have been shown to be 90% and more than 90%, respectively. Owing to such a high homology among the amino acid sequences of Arabidopsis actin isoforms, little is known regarding the differences in their biochemical properties. Therefore, to investigate the individual biochemical properties of these isoforms, each Arabidopsis isoforms was purified by expression as actin–thymosin-β–His-tag fusion proteins in *Dictyostelium discoideum* cells (Kijima et al. 2016). Thymosin-β prevented harmful copolymerisation with Dictyostelium actins in cells. After purifying actin–thymosin-β–His-tag proteins using Ni-affinity column chromatography, the thymosin-β–His-tag was deleted by the action of a specific protease.

Interestingly, the biochemical properties of purified Arabidopsis actin isoforms were different. First, the binding properties with phalloidin differed among the actin isoforms. Phalloidin could strongly bind to AtACT1 and AtACT11 filaments; it could bind only weakly to AtACT7 filaments and hardly at all to AtACT2 filaments. Second, the polymerisation rates of AtACT2 and AtACT7 (expressed in vegetative organs) were 2–3 times faster than those of AtACT1 and AtACT11 (expressed in reproductive organs). Third, the phosphate release rate from actin filaments differed among Arabidopsis actin isoforms. Finally, the interactions to profilin were different among actin isoforms. These in vitro data suggest that Arabidopsis actin isoforms play distinct roles in vivo. No conventional actin-staining methods, such as those using rhodamine–phalloidin, indirect immune fluorescence or actin markers fused with fluorescent protein, can distinguish between these actin isoforms. Therefore, to visualise different subcellular localisations of Arabidopsis actin isoforms, each actin isoform was fused with fluorescent proteins via a long linker peptide (6XGSS) and expressed in plant cells (Kijima et al. 2018). Interestingly, AtACT2 and AtACT7 polymerised as different types of filaments in leaf mesophyll cells of *Nicotiana benthamiana*. These results suggest that different actin isoforms form different actin structures or networks in plant cells.

4.3 Diversity of Plant Myosin XIs

Angiosperms possess over 10 myosin XI isoforms in their genomes, while algae, mosses, and ferns possess relatively few myosin XI isoforms (Fig. 4.1). Myosin XIs have been implicated to play a role in the generation of cytoplasmic streaming as well as in more complex biological processes in vascular plants. In contrast, the number of myosin VIII isoforms has not diversified as much (Fig. 4.1). Thus, the physiological functions of myosin VIII are likely to be relatively conserved, whereas those of myosin XI have specifically diversified to regulate highly developed systems during plant evolution. To clarify myosin XI diversity, tissue-specific expression patterns, velocities, and enzymatic activities, all 13 myosin XI isoforms in *Arabidopsis thaliana* have been studied in detail (Haraguchi et al. 2018). Promoter-reporter assays have revealed that expression patterns of Arabidopsis myosin XIs were divided into three categories: ubiquitous expression pattern, AtXI-1, -2, -B, and -K; pollen-specific expression pattern, AtXI-A, -C, -D, -E, and -J; and other expression patterns, AtXI-F, -G, -H, and -I (Fig. 4.2). Moreover, recombinant Arabidopsis myosin XIs expressed in insect cells have demonstrated diverse motor activities. The in vitro velocities of Arabidopsis myosin XIs were divided into three groups: the low-velocity group ($0.1~\mu m~s^{-1}$), At XI-I; the medium-velocity group ($5–7~\mu m~s^{-1}$), AtXI-1, -2, -B, -K, -H, and J; and the high-velocity group ($12–23~\mu m~s^{-1}$), AtXI-A, -C, -D, -E, -F, and -G. Interestingly, this classification based on velocity appeared to be crudely associated with the tissue-specific expression patterns of these myosin XIs (Fig. 4.2). In general, the ubiquitously expressed Arabidopsis myosin XIs, which are the motive force during cytoplasmic streaming, belong to the medium-velocity group, while the pollen-specific Arabidopsis myosin XIs belong to the high-velocity group, with only AtXI-I belonging to the low-velocity group. These results suggest that the molecular functions of Arabidopsis myosin XIs have diversified for their specific tasks *in planta*.

Phylogenetic analysis using motor domain sequences have revealed that myosin XIs are generally classified into 8 lineages, namely Myo11-A, -B, -C, -D, -E, -F, -G, and -H (Madison and Nebenführ. 2013). To investigate the association between phylogenetic classification and functional diversity of myosin XIs, we generated a phylogenetic tree of various plant myosin XIs and compared their phylogeny with the corresponding expression patterns and velocities (Fig. 4.3). There was no association between the lineages and expression patterns, although there was some association between the lineages and velocities. Low-, middle- and high-velocity at myosin XIs tend to exist in different lineage. Anyway, the correlation between sequence similarity and myosin XI velocity should be confirmed by more in vitro data obtained from other plant species.

Furthermore, biochemical approach demonstrated variation in actin–myosin XI interactions. Individual Arabidopsis myosin XIs exhibited different motile activities when in association with different Arabidopsis actin isoforms (Rula et al. 2018). Two Arabidopsis myosin isoforms (AtXI-2 and -B) and three Arabidopsis actin isoforms (AtACT1, 2, and 7) were used for motile activity studies. The velocity of the ubiquitously expressed AtXI-2 was higher in association with AtACT2 and AtACT7 (expressed in vegetative organs) than that in association with AtACT1 (expressed in reproductive organs). Conversely, the velocity of the ubiquitously expressed AtXI-B was higher when in association with AtACT1 (expressed in the reproductive organs) than that in association with AtACT2 (expressed in the vegetative organs). The expression patterns and intensities of Arabidopsis actin isoforms and Arabidopsis myosin XIs are summarised in Fig. 4.2.

4.4 Conclusions

Recent studies have demonstrated that land plants have developed unique actin–myosin systems not only for the generation of cytoplasmic streaming but also for regulation of various biological phenomena (Duan and Tominaga 2018; Tominaga

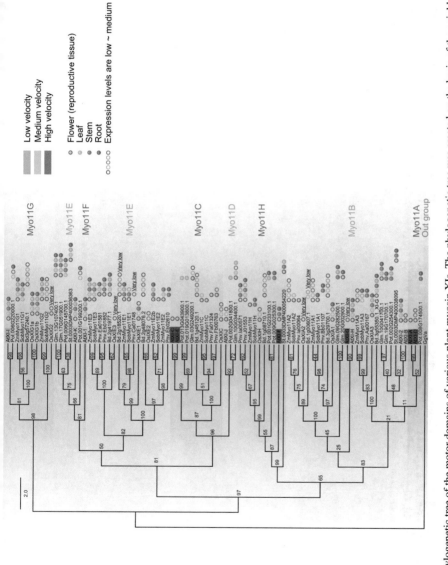

Fig. 4.3 Phylogenetic tree of the motor domains of various plant myosin XIs. The phylogenetic tree was created on the basis of the neighbour-joining method using Clustal X (Thompson et al. 1997). Amino acid sequence and expression data were obtained from Phytozome (https://phytozome.jgi.doe.gov/pz/portal.html). Numbers indicated percentage bootstrap value of 10,000 iterations. Each colour represents lineages determined by Madison and Nebenführ. (2013).

Fig. 4.3 (continued) Expression sites are coloured according to organ, namely reproductive organ (red), leaf (green), stem (blue), and root (purple). Expression levels are shown by colour intensity; the stronger the colour intensity, the higher the expression. GgVa (*Gallus gallus* myosin Va) was used as an outgroup. *Arabidopsis thaliana* (At), *Oryza sativa* (Os), *Brachypodium distachyon* (Bd), *Zea mays* (Zm), *Glycine max* (Glm), *Panicum virgatum* (Pnv), *Sorghum bicolor* (Sob), *Solanum tuberosum* (St) and *Populus trichocarpa* (Pot)

and Ito 2015). Figure 4.1 presents the number of actin, myosin VIII, and XI isoforms in Charophytes, Bryophytes, Lycopodiophytes, Gymnosperms, Monocotyledons, and Eudicotyledons. These data clearly indicate that myosin XI, in particular, diversified during plant evolution to angiosperms. Nearly half of the actin and myosin isoforms present in Arabidopsis were expressed in pollen, which has evolved as part of the reproductive system in angiosperms suitable for the terrestrial environment; this suggests that the diversification of actin and myosin isoforms may have been necessary for the evolution of the fertilisation system from 'sperm' to 'pollen tube'. Pollen tube growth is an important initial event during fertilisation, which is mediated by vesicle transport towards the tip of pollen tube (tip growth), involving actin–myosin XI interactions (Cai et al. 2015). Furthermore, the direction of pollen tube growth in the style is highly regulated by peptide signals (LURE) secreted by the synergid cells and receptors (RPK) at the pollen tube tip (pollen tube guidance) (Okuda et al. 2009; Takeuchi and Higashiyama 2016). However, the intracellular mechanism controlling the direction of pollen tube growth remains unclear. Here, we propose a hypothesis based on an 'affinity-selective transport model' via the specific selection of vesicles depending on the affinity to the actin–myosin complex (Fig. 4.4). In pollen tubes, three types of actin structures are present (Cai et al. 2015): (1) longitudinal thick actin bundles in the shank; (2) short dynamic actin bundles in the subapical region (actin fringes); and (3) short highly dynamic actin filaments at the tip. As Kijima et al. have found in epidermal cells (2018), each actin structure in the pollen tube may be composed of different actin isoforms (as illustrated by different colours in Fig. 4.4) having various affinities to different myosin XI isoforms. Both organelles and vesicles are transported by specific myosin XI isoforms along the long actin bundles towards the tip of the pollen tube. At the subapical region, vesicles are filtrated and selected not only by the physical barrier but also by the affinity between specific actin and myosin XI isoforms through the actin fringe. Selectively transported vesicles detach from actin at the apical dome via calcium-induced myosin XI released from actin (Tominaga et al. 2012; Tominaga and Nakano 2012). Finally, the vesicles fuse with the plasma membrane at the tip, and cell wall precursors and cellulose synthetase are introduced to the extracellular space. Myosin XI is recycled in an inactive folded conformation or through retrograde membrane transport. In this model, the actin bundles in the fringe could regulate the direction of pollen tube growth by acting as a junction to control the number of vesicles being transported, that is, extracellular signal (LURE) induced reconstitution of actin bundles in the fringe, consequently it caused uneven distributions of actin isoforms in the fringe (as illustrated by orange and purple actin bundles in Fig. 4.4). The reconstitution of these actin bundles may be triggered through the PRK receptor-mediated ROP signalling (Takeuchi and Higashiyama 2016). Subsequently, the different actin isoform distributions in the fringe mediate uneven passage of the vesicles depending on the actin–myosin XI affinity. Finally, biased vesicle transport and fusion at the apical region leads to biased tip growth, directed by the signal. To confirm whether the 'affinity-selective transport model' is correct, simultaneous visualisation and real-time imaging of each actin and myosin isoforms are imperative.

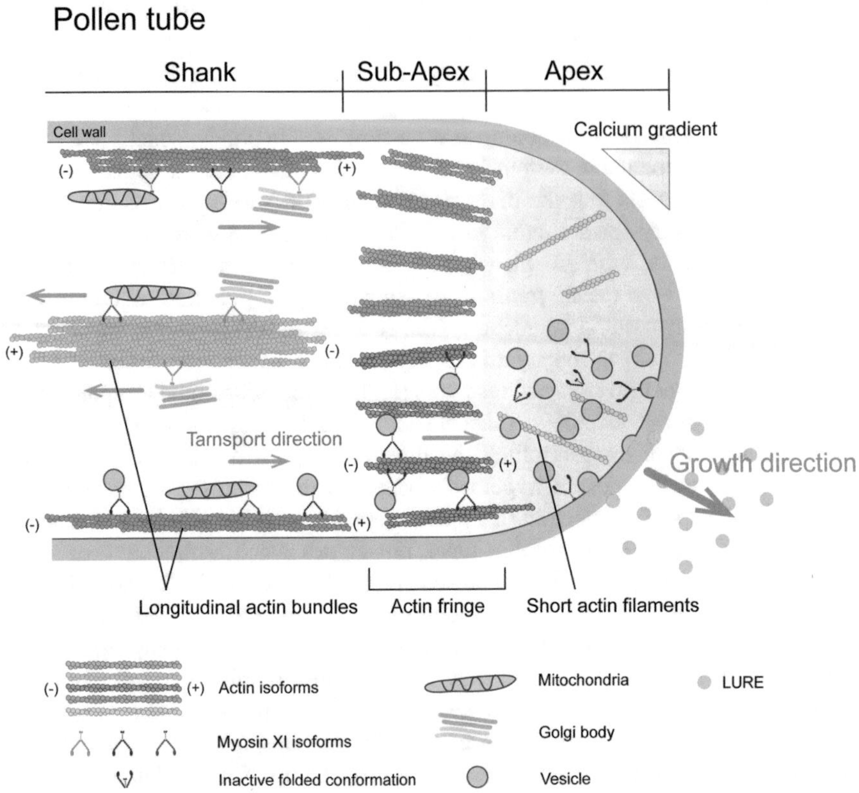

Pollen tube

Fig. 4.4 The proposed affinity-selective transport model in the pollen tube. In the pollen tube, there are three types of actin structure: (1) longitudinal thick actin bundles in the shank (red and green), (2) short dynamic actin bundles in the subapical region (actin fringe; orange and purple), and (3) short highly dynamic actin filaments at the tip (blue). Each of the three actin structures may be composed of different actin isoforms (as illustrated by the different colours). In this model, vesicle transport is selectively regulated by affinity between actin and myosin XI isoforms

As summarised in this review, the plant actin–myosin system has diversified from the molecular to the tissue level to a greater extent than expected. This diversity may be critical for the upgrade of biological functions during plant evolution. To reveal the mechanisms regulated by the plant-specific actin–myosin systems, an integrated approach, including gene knockout analyses, gene modification, live-cell imaging, and biochemical–mechanical analyses, is warranted.

Acknowledgements The authors would like to thank the Japan Society for the Promotion of Science [KAKENHI 24658002, 26440131, and 15H01309 (to K.I.), 20001009, 23770060, and 25221103 (to M.T.)] and the Japan Science and Technology Agency, ALCA, [JPMJAL1401 (to K. I., T.H., Z.D., and M.T.)] for support.

References

Avisar D, Prokhnevsky AI, Dolja VV (2008) Class VIII myosins are required for plasmodesmatal localization of a closterovirus Hsp70 homolog. J Virol 82:2836–2843. https://doi.org/10.1128/JVI.02246-07

Baluska F, Cvrckova F, Kendrick-Jones J, Volkmann D (2001) Sink plasmodesmata as gateways for phloem unloading. Myosin VIII and calreticulin as molecular determinants of sink strength? Plant Physiol 126:39–46

Cai G, Parrotta L, Cresti M (2015) Organelle trafficking, the cytoskeleton, and pollen tube growth. J Integr Plant Biol 57:63–78. https://doi.org/10.1111/jipb.12289

Duan Z, Tominaga M (2018) Actin-myosin XI: an intracellular control network in plants. Biochem Biophys Res Commun 506:403–408. https://doi.org/10.1016/j.bbrc.2017.12.169

El-Mezgueldi M, Bagshaw CR (2008) The myosin family: biochemical and kinetic properties. In: Myosins. Springer, Berlin, pp 55–93

Golomb L, Abu-Abied M, Belausov E, Sadot E (2008) Different subcellular localizations and functions of Arabidopsis myosin VIII. BMC Plant Biol 8:3. https://doi.org/10.1186/1471-2229-8-3

Gunning PW, Ghoshdastider U, Whitaker S, Popp D, Robinson RC (2015) The evolution of compositionally and functionally distinct actin filaments. J Cell Sci 128:2009–2019. https://doi.org/10.1242/jcs.165563

Haraguchi T, Tominaga M, Matsumoto R, Sato K, Nakano A, Yamamoto K, Ito K (2014) Molecular characterization and subcellular localization of Arabidopsis class VIII myosin, ATM1. J Biol Chem 289:12343–12355. https://doi.org/10.1074/jbc.M113.521716

Haraguchi T et al (2018) Functional diversity of class XI myosins in Arabidopsis thaliana. Plant Cell Physiol 59:2268–2277. https://doi.org/10.1093/pcp/pcy147

Ito K et al (2003) Recombinant motor domain constructs of Chara corallina myosin display fast motility and high ATPase activity. Biochem Biophys Res Commun 312:958–964

Kandasamy MK, McKinney EC, Meagher RB (2009) A single vegetative actin isovariant overexpressed under the control of multiple regulatory sequences is sufficient for normal Arabidopsis development. Plant Cell 21:701–718. https://doi.org/10.1105/tpc.108.061960

Kato T, Morita MT, Tasaka M (2010) Defects in dynamics and functions of actin filament in Arabidopsis caused by the dominant-negative actin fiz1-induced fragmentation of actin filament. Plant Cell Physiol 51:333–338. https://doi.org/10.1093/pcp/pcp189

Kijima ST, Hirose K, Kong SG, Wada M, Uyeda TQ (2016) Distinct biochemical properties of Arabidopsis thaliana actin isoforms. Plant Cell Physiol 57:46–56. https://doi.org/10.1093/pcp/pcv176

Kijima ST, Staiger CJ, Katoh K, Nagasaki A, Ito K, Uyeda TQP (2018) Arabidopsis vegetative actin isoforms, AtACT2 and AtACT7, generate distinct filament arrays in living plant cells. Sci Rep 8:4381. https://doi.org/10.1038/s41598-018-22707-w

Kollmar M, Muhlhausen S (2017) Myosin repertoire expansion coincides with eukaryotic diversification in the Mesoproterozoic era. BMC Evol Biol 17:211. https://doi.org/10.1186/s12862-017-1056-2

Lanza M et al (2012) Role of actin cytoskeleton in brassinosteroid signaling and in its integration with the auxin response in plants. Dev Cell 22:1275–1285. https://doi.org/10.1016/j.devcel.2012.04.008

Madison SL, Nebenführ A (2013) Understanding myosin functions in plants: are we there yet? Curr Opin Plant Biol 16(6):710–717. https://doi.org/10.1016/j.pbi.2013.10.004

Madison SL, Buchanan ML, Glass JD, McClain TF, Park E, Nebenfuhr A (2015) Class XI myosins move specific organelles in pollen tubes and are required for normal fertility and pollen tube growth in Arabidopsis. Plant Physiol 169:1946–1960. https://doi.org/10.1104/pp.15.01161

McDowell JM, Huang S, McKinney EC, An YQ, Meagher RB (1996) Structure and evolution of the actin gene family in Arabidopsis thaliana. Genetics 142:587–602

Meagher RB, McKinney EC, Vitale AV (1999) The evolution of new structures: clues from plant cytoskeletal genes. Trends Genet 15:278–284

Muhlhausen S, Kollmar M (2013) Whole-genome duplication events in plant evolution reconstructed and predicted using myosin motor proteins. BMC Evol Biol 13:202. https://doi.org/10.1186/1471-2148-13-202

Nishiyama T et al (2018) The *Chara* genome: secondary complexity and implications for plant terrestrialization. Cell 174:448–464.e24. https://doi.org/10.1016/j.cell.2018.06.033

Ojangu EL, Tanner K, Pata P, Jarve K, Holweg CL, Truve E, Paves H (2012) Myosins XI-K, XI-1, and XI-2 are required for development of pavement cells, trichomes, and stigmatic papillae in *Arabidopsis*. BMC Plant Biol 12:81. https://doi.org/10.1186/1471-2229-12-81

Okamoto K et al (2015) Regulation of organ straightening and plant posture by an actin-myosin XI cytoskeleton. Nat Plant 1:15031

Okuda S et al (2009) Defensin-like polypeptide LUREs are pollen tube attractants secreted from synergid cells. Nature 458:357–361. https://doi.org/10.1038/nature07882

Peremyslov VV, Prokhnevsky AI, Avisar D, Dolja VV (2008) Two class XI myosins function in organelle trafficking and root hair development in Arabidopsis. Plant Physiol 146:1109–1116. https://doi.org/10.1104/pp.107.113654

Peremyslov VV, Prokhnevsky AI, Dolja VV (2010) Class XI myosins are required for development, cell expansion, and F-actin organization in *Arabidopsis*. Plant Cell 22:1883–1897. https://doi.org/10.1105/tpc.110.076315

Prokhnevsky AI, Peremyslov VV, Dolja VV (2008) Overlapping functions of the four class XI myosins in *Arabidopsis* growth, root hair elongation, and organelle motility. Proc Natl Acad Sci USA 105:19744–19749. https://doi.org/10.1073/pnas.0810730105

Reddy ASN (2001) Molecular motors and their functions in plants. Int Rev Cytol 204:97–178

Reichelt S, Knight AE, Hodge TP, Baluska F, Samaj J, Volkmann D, Kendrick-Jones J (1999) Characterization of the unconventional myosin VIII in plant cells and its localization at the post-cytokinetic cell wall. Plant J 19:555–567

Rula S et al (2018) Measurement of enzymatic and motile activities of *Arabidopsis* myosins by using *Arabidopsis* actins. Biochem Biophys Res Commun 495:2145–2151. https://doi.org/10.1016/j.bbrc.2017.12.071

Sattarzadeh A, Franzen R, Schmelzer E (2008) The *Arabidopsis* class VIII myosin ATM2 is involved in endocytosis. Cell Motil Cytoskeleton 65:457–468. https://doi.org/10.1002/cm.20271

Shimmen T, Yokota E (2004) Cytoplasmic streaming in plants. Curr Opin Cell Biol 16:68–72. https://doi.org/10.1016/j.ceb.2003.11.009

Slajcherova K, Fiserova J, Fischer L, Schwarzerova K (2012) Multiple actin isotypes in plants: diverse genes for diverse roles? Front Plant Sci 3:226. https://doi.org/10.3389/fpls.2012.00226

Takeuchi H, Higashiyama T (2016) Tip-localized receptors control pollen tube growth and LURE sensing in *Arabidopsis*. Nature 531:245–248. https://doi.org/10.1038/nature17413

Tamura K et al (2013) Myosin XI-i links the nuclear membrane to the cytoskeleton to control nuclear movement and shape in *Arabidopsis*. Curr Biol 23:1776–1781. https://doi.org/10.1016/j.cub.2013.07.035

Thompson JD, Gibson TJ, Plewniak F, Jeanmougin F, Higgins DG (1997) The CLUSTAL_X windows interface: flexible strategies for multiple sequence alignment aided by quality analysis tools. Nucleic Acids Res 25:4876–4882

Tominaga M, Ito K (2015) The molecular mechanism and physiological role of cytoplasmic streaming. Curr Opin Plant Biol 27:104–110. https://doi.org/10.1016/j.pbi.2015.06.017

Tominaga M, Nakano A (2012) Plant-specific myosin XI, a molecular perspective. Front Plant Sci 3:211. https://doi.org/10.3389/fpls.2012.00211

Tominaga M et al (2003) Higher plant myosin XI moves processively on actin with 35 nm steps at high velocity. EMBO J 22:1263–1272. https://doi.org/10.1093/emboj/cdg130

Tominaga M, Kojima H, Yokota E, Nakamori R, Anson M, Shimmen T, Oiwa K (2012) The calcium-induced mechanical change in the neck domain alters the activity of plant myosin XI. J Biol Chem 287:30711–30718

Tominaga M et al (2013) Cytoplasmic streaming velocity as a plant size determinant. Dev Cell 27:345–352. https://doi.org/10.1016/j.devcel.2013.10.005

Ueda H et al (2010) Myosin-dependent endoplasmic reticulum motility and F-actin organization in plant cells. Proc Natl Acad Sci USA 107:6894–6899. https://doi.org/10.1073/pnas.0911482107

Yamamoto K, Kikuyama M, Sutoh-Yamamoto N, Kamitsubo E (1994) Purification of actin based motor protein from *Chara corallina*. Proc Jpn Acad 70:175–180

Yamamoto K, Hamada S, Kashiyama T (1999) Myosins from plants. Cell Mol Life Sci 56:227–232

Yang L, Qin L, Liu G, Peremyslov VV, Dolja VV, Wei Y (2014) Myosins XI modulate host cellular responses and penetration resistance to fungal pathogens. Proc Natl Acad Sci USA 111:13996–14001. https://doi.org/10.1073/pnas.1405292111

Yokota E, Shimmen T (1994) Isolation and characterization of plant myosin from pollen tubes of lily. Protoplasma 177:153–162

Chapter 5
Actin Cytoskeleton and Action Potentials: Forgotten Connections

F. Baluška and S. Mancuso

Abstract Actin cytoskeleton was discovered some 70 years ago, and it is well known to be responsible for cellular transport phenomena and contractilities, with animal muscles representing the most obvious example. This ancient cytoskeletal system is present in all eukaryotic cells, responsible for all kinds of intracellular motilities. For example, the synaptic vesicle recycling also relies on the actin cytoskeleton, which supports all types of membranes structurally and functionally. Action potentials are fundamental for the long-distance signaling in both animals and plants. Although it is not generally appreciated, action potentials are mechanistically and functionally interlinked with the actin cytoskeleton associated with membranes. In both animals and plants, the inherent bioelectricity of membranes is closely linked with the actin cytoskeleton. Despite the fundamental importance of this phenomenon, it remains to be under-investigated, and future studies will be needed to illuminate the elusive electrochemical and bioelectric nature of cellular life.

5.1 Introduction

Although immobile plants are generally not considered relevant for studies on motilities based on the actin cytoskeleton and long-distance bioelectric and electrochemical communication via action potentials, the fact is that plants, especially the large algal cells of Characeae, served as suitable model objects for early studies on intracellular motilities and action potentials (Wayne 1993; Stahlberg 2006; Beilby 2007). Intriguingly, already in 1901, Bohumil Němec used a unique histochemical method which visualized filamentous structures and cables closely resembling F-actin cables in cells of root apices (Němec 1901; see the Fig. 1 in Baluška and

F. Baluška (✉)
University of Bonn, IZMB, Bonn, Germany
e-mail: baluska@uni-bonn.de

S. Mancuso
University of Firenze, LINV, Florence, Italy
e-mail: stefano.mancuso@unifi.it

© Springer Nature Switzerland AG 2019
V. P. Sahi, F. Baluška (eds.), *The Cytoskeleton*, Plant Cell Monographs 24,
https://doi.org/10.1007/978-3-030-33528-1_5

Hlavačka 2005). Němec considered these filamentous structures relevant to transmission of sensory excitations. Currently, there is an active and vital field studying the plant cytoskeleton, including the actin cytoskeleton, and our understanding of the roles of F-actin in plant cells is expanding significantly (Volkmann and Baluška 1999; Staiger et al. 2000; Wang and Hussey 2015; Wang et al. 2017a, b).

Importantly, as in animal cells and neurons, F-actin is important for endocytosis and relevant for cell-to-cell communication (Baluška et al. 2000, 2005, 2009, 2010; Baluška and Hlavačka 2005; Baluška and Mancuso 2013). Intriguingly, the action potentials were found to block, fastly but transiently, the actin-based cytoplasmic streaming. This important connection between bioelectricity and cell biology was discovered very early (Hörmann 1898; Blinks et al. 1929; Sibaoka and Oda 1956) during the first characterization of the electrochemical nature of action potentials in large cells of Characean algae as well as giant squid axons (Blinks et al. 1929; Umrath 1932; Osterhout 1936; Cole and Curtis 1938, 1939; Keynes 1958; Tazawa and Kishimoto 1968). Unfortunately, this important connection between the electrophysiology and cell biology is still only poorly understood. Even worse, this topic is even not included in the agenda of contemporary experimental studies in both the medical and botanical fields. Intriguingly and relevantly in this respect, diverse anesthetics block action potentials in both animal and plant cells and also inhibit cytoplasmic streaming (Ewart 1902, 1903; Osterhout 1936; Fink and Kish 1976; Lavoie et al. 1989; García-Sierra and Frixione 1993).

5.2 Actin Cytoskeleton in Plants

The actin cytoskeleton is ancient structural system of all eukaryotic organisms that plays central roles in organization of intracellular trafficking of vesicles and organelles in close interactions with signaling and communication (Volkmann and Baluška 1999; Staiger et al. 2000; Li et al. 2015; Wang and Hussey 2015; Wang et al. 2017a, b). Actin filaments (F-actin) are polymerized from globular actin monomers (G-actin) and dynamically assembled (Li et al. 2015) into diverse F-actin arrays composed of bundles and meshworks (Volkmann and Baluška 1999). Besides intracellular trafficking, actin cytoskeleton also supports the integrity of the plasma membrane via exocytosis and endocytosis processes (Baluška et al. 2004; Šamaj et al. 2005; Sattarzadeh et al. 2008), the gating of the plant-specific cytoplasmic cell–cell channels, the plasmodesmata (Radford and White 1998; Reichelt et al. 1999; Baluška et al. 1999, 2001a; Volkmann et al. 2003; White and Barton 2011; Diao et al. 2018), as well as the onset of rapid cell elongation of root cells (Baluška et al. 1997, 2001b; Takatsuka et al. 2018).

Although actin was discovered and characterized biochemically by Brunó Ferenc Straub and Albert Szent-Györgyi in 1942 (Straub 1942; Szent-Györgyi 1942, 1943, 2004), its first intracellular visualization succeeded some twenty years later using electron microscopy (for plant cells, see Palevitz et al. 1974; Palevitz and Hepler 1975; Kersey et al. 1976). Another twenty years later, the indirect

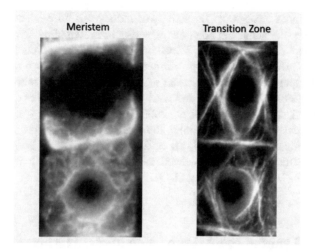

Fig. 5.1 Actin cytoskeleton visualized with anti-actin antibody applied on Steedman's wax section of maize root apex. Left side: In mitotically dividing cells of the root apical meristem, F-actin arrays are organized in the form of dense meshworks suspending the centrally positioned interphase nucleus (lower cell) of the meristem. During mitotic division (upper cell), F-actin arrays accumulate at the future cross-walls (end-poles) which face and polarize the mitotic spindle poles, and F-actin is depleted from the area of mitotic spindle. Right side: Postmitotic cells of the root apex transition zone show dense F-actin arrays at the cross-walls (end-poles) which organize prominent F-actin bundles enclosing the still centrally positioned nucleus. This unique F-actin configuration is specific for cells of the transition zone. See Voigt et al. (2005a) for similar situation in living *Arabidopsis* roots expressing the F-actin marker GFP-FABD2 construct. These images are taken from Baluška et al. (1997)

immunofluorescence technique allowed visualization of the actin cytoskeleton inside cells, using specific actin antibodies. This technique entered also plant cell biology studies in the eighties of the last century (Clayton and Lloyd 1985; Seagull et al. 1987; Traas et al. 1987). However, these early studies used plant protoplasts and the first indirect immunofluorescence localization of the actin cytoskeleton in intact cells organized within plant tissues was succeeded only in 1997 using the Steedman's wax technique (Baluška et al. 1997). This technique yielded unexpected results in maize root apex cells (Fig. 5.1): (1) abundant F-actin meshworks at the nongrowing cross-walls (end-poles) and (2) prominent F-actin cables enclosing centrally positioned nuclei (Baluška et al. 1997). Intriguingly, this unique F-actin arrangement was reported by Bohumil Němec already in 1901 using histochemical reactions (Němec 1901; see Fig. 1 in Baluška and Hlavačka 2005; Fig. 5.1). Němec speculated that these fibrils are involved in transmission of sensory information from the apical sensory root cap cells (specialized for graviperception) to the basal motoric and rapidly elongating cells (Němec 1901; Baluška et al. 2006, 2009). Two years later, these F-actin meshworks at the nongrowing cross-walls were associated with the plant-specific myosin VIII (Reichelt et al. 1999; Baluška et al. 2001a; Volkmann et al. 2003), which supports both active endocytosis and endocytic vesicle recycling (Baluška et al. 2000, 2001a, 2004, 2005; Šamaj et al. 2005; Sattarzadeh et al. 2008).

At the root apex cross-walls, the activity of endocytosis fully balances exocytosis, resulting in the very active but nongrowing end-poles, resembling in many aspects of neuronal synapses (Baluška 2010; Baluška et al. 2005, 2006, 2009, 2010; Baluška and Mancuso 2013). These plant-specific synapses emerge to be active in the cell-to-cell polar transport of auxin via vesicular secretion and endocytic recycling (Baluška et al. 2005, 2008, 2018; Baluška and Mancuso 2013; Mancuso et al. 2005, 2007; Mettbach et al. 2017; Schlicht et al. 2006; Zhu et al. 2016). Cell-to-cell auxin transport is sensitive to specific inhibitors of exocytosis, vesicle recycling, and F-actin dynamics (Mancuso et al. 2005, 2007; Schlicht et al. 2006; Dhonukshe et al. 2008; Zhu et al. 2016), as well as to mutants affecting these processes (Mancuso et al. 2005, 2007; Zhu et al. 2016).

5.3   Two Types of Plant Action Potentials: Plasma Membrane Versus Vacuolar Membranes

The plant protoplasm is sandwiched between two apoplastic compartments: plasma membrane faces toward the cell wall, whereas the tonoplast membrane surrounds the vacuolar contents. Both these membranes are excitable and support action potentials (Findlay and Hope 1964; Wayne 1993; Beilby 1984, 2007, 2016; Shepherd et al. 1998; Hedrich 2012).

The actin cytoskeleton is inherently associated not only with the plasma membrane but also with the vacuolar tonoplast membranes (Gao et al. 2009; Li et al. 2013; Scheuring et al. 2016). This tonoplast associated F-actin can be expected to be relevant to the onset of rapid cell elongation which is based on the explosive buildup of large cellular vacuoles at the beginning of the rapid cell elongation zone (Baluška et al. 1997, 2001b; Takatsuka et al. 2018; Dünser et al. 2019). Importantly in this respect, the plants enjoy two different types of action potentials. The classical one is based on the plasma membrane ion fluxes, while the vacuolar action potential describes transmembrane ion fluxes across the tonoplast membrane (Findlay and Hope 1964; Kikuyama and Tazawa 1976; Kikuyama 1986, 2001; Heidecker et al. 2003; Kikuyama and Shimmen 1997). Interestingly, the integration of the excitabilities of the plasma membrane and the tonoplast is still poorly understood (Wayne 1993; Kikuyama 2001; Hedrich 2012). Besides calcium spikes (Kikuyama 1986; Wayne 1993), vesicular trafficking between the cell compartments, especially endocytic recycling vesicles, may be relevant in this respect too (Shepherd et al. 2004). Both the plasma membrane and the tonoplast are equipped with electrogenic pumps driving transport of ions (either way) and protons out of the cytoplasm (Wayne 1993; Beilby 2007; Hedrich 2012); either into the cell walls or into vacuoles and endosomes. The resting plasma membrane potential of plant cells is about -110 to -250 mV (Wayne 1993; Roelfsema et al. 2001; Beilby 2007; Hedrich 2012; Pedersen et al. 2012), while those of tonoplast membranes range from 0 to -30 mV (Wayne 1993; Walker et al. 2006; Hedrich 2012). Plasma membrane H^+-ATPases generate very steep H^+ gradients which allow potentials up to 250 mV (Hirsch et al. 1989; Pedersen et al. 2012). In contrast to more complex plant cells, excitable animal and human cells, including neurons, have lower resting plasma membrane potentials: in a range from -65 to $-80\,mV$ (Bean 2007; Bezanilla 2006, 2008). Moreover, all neurons also lack vacuoles as well as vacuolar action potentials.

In plants, vacuolar action potentials are downstream of the plasma membrane action potentials. The latter induce an increase of the cytoplasmic calcium levels (Beilby 1984; Shepherd et al. 1998; Wacke and Thiel 2001; Wacke et al. 2003) which activate chloride channels on the vacuolar tonoplast membranes, allowing chloride ions to move out of the vacuoles into the cytoplasm (Kikuyama 1986, 2001; Shimmen and Nishikawa 1988; Kikuyama and Shimmen 1997; Wayne 1993; Beilby 2007, 2016). It is important to note that the action potential mechanisms on both membranes are used by the salt-tolerant Characeae to rapidly dump K^+ and Cl^- out of the cell in the hypotonic turgor regulation (Beilby and Shepherd 1996). It might be expected that a similar process is important also in root cells of salt-tolerant land plants.

In Characeae *Nitella* and *Nitellopsis*, the vacuolar action potentials were sometimes observed without accompanying plasma membrane action potential (Kisnieriene et al. 2019). This implicates an independent nature of both kinds of action potentials in plants, which are typically integrated into the whole cell electrophysiology, perhaps via the cytoplasmic calcium transients. Thiel–Beilby model of plant action potentials suggests that the threshold depolarizations of the plasma membrane are translated into calcium spikes due to its release from the internal stores (Wacke and Thiel 2001; Wacke et al. 2003; Beilby and Al Khazaaly 2016; Kisnieriene et al. 2019). The updated Thiel–Beilby model implicates diacylglycerol (DAG) in the role of critical signaling molecule mediating impacts of calcium spikes on generation of plant action potentials (Beilby and Al Khazaaly 2016). Interestingly in this respect, plant DAG emerges as important signaling molecule regulating vesicle trafficking together with phosphatidic acid (PA) in plant rapid responses to environmental stimuli (Arisz et al. 2009; Dong et al. 2012). Recent in vivo analysis revealed that DAG is produced at the cytoplasmic leaflets of membranes in plant cells; it is also present within the cytoplasm and nuclei, and abundant at the plasma membrane of the root apex transition zone cells (Vermeer et al. 2017), which are active in the electric spike activities (Masi et al. 2009, 2015) as well as in the synaptic cell–cell communication activities (Baluška 2010; Baluška et al. 2005, 2006, 2009, 2010; Baluška and Mancuso 2013). Finally, both PA and DAG are critical signaling molecules in the control of actin polymerization and F-actin dynamics (Li et al. 2012a; Li and Staiger 2018; Pleskot et al. 2012, 2013, 2014).

5.4 Action Potentials Evolved from Ancient Actin-Based Membrane Repair Processes

In the evolution of unicellular and multicellular organisms, action potentials evolved from ancient mechanisms safeguarding the structural and functional integrity of membranes and cells. Also to keep low cytoplasmic Ca^{2+} not to interfere with phosphate-based energetic metabolism, the use of Ca^{2+} from internal stores can be fastly resequestered (Brunet and Arendt 2016). This theory was proposed already in 1982 by Andrew Goldsworthy for plants, and later also for neurons of animals and humans (Goldsworthy 1983; Steinhardt et al. 1994; Togo et al. 1999; Andrews et al. 2014; Brunet and Arendt 2016). The central idea behind this *neuronal* theory of the plasma membrane repair is that there is an efficient homeostatic system devoted to preserve the plasma membrane integrity. This hypothesis is based on the voltage-gated ion channels and the membrane-associated actin cytoskeleton driving calcium controlled endocytic vesicle recycling (Steinhardt et al. 1994; Togo et al. 1999). Cells are very effective in self-repairing throughout the evolutionary morphospace, from the unicellular protozoa, via algal (Menzel 1988; Foissner and Wasteneys 2012) and higher plant cells, up to animal and human neurons (Tang and Marshall 2017). Cells use their calcium channels and the signaling pathways impinging on specialized proteins such as synaptotagmins and annexins. These allow tight couplings between ionic/electrochemical phenomena at the excitable membranes and the dynamic actin cytoskeleton in the regulation of vesicle trafficking, exocytosis, and endocytosis (Baluška and Wan 2012; Baluška et al. 2005, 2009; Berghöfer et al. 2009; Brunet and Arendt 2016). We have discovered plant synaptotagmins which have the features of both classical neuronal synaptotagmins and the extended synaptotagmins (Craxton 2007; Schapire et al. 2008; Saheki and De Camilli 2017). These plant-specific synaptotagmins are essential for the plasma membrane repair under diverse stress situations (Schapire et al. 2008, 2009; Yamazaki et al. 2008, 2010; Pérez-Sancho et al. 2015; Siao et al. 2016; Wang et al. 2017a, b, 2018). Annexins are, similarly to synaptotagmins, calcium-induced phospholipid-binding membrane proteins which, besides their active roles in membrane repair via vesicle trafficking, also act as calcium channels in plant cells (Clark et al. 2012; Davies 2014; Richards et al. 2014). Annexins are also actively involved in plasma membrane repair processes in eukaryotic cells (Boye and Nylandsted 2016; Boye et al. 2017; Simonsen et al. 2019).

5.5 Action Potentials Induce Structural Changes in Cells Related to Actin Cytoskeleton

Since the very early studies of action potentials in large algal cells and squid neurons, it was obvious that these rapid depolarizations of the plasma membrane are associated with rapid but brief changes of cellular structures including inhibition of cytoplasmic streaming and shortening of excised cells (Hill 1941; Kishimoto and Akabori 1959; Kishimoto and Ohkawa 1966; Barry 1970a, b; Cohen 1973; Yoshioka and Takenaka 1979). Moreover, an increased pressure was measured during action potential in nerve fibers from crab claws and lobster legs (Iwasa et al. 1980; Tasaki et al. 1980). During the action potential in neurons, it is well known that there is a sudden but brief reversal of the plasma membrane voltage, when it changes from its resting value of about -70 mV (negative inside) to about $+30$ mV in less than 1 ms, and then returns back to the original resting value in just a few milliseconds (Bezanilla 2006, 2008). In large algal cells of Characeae, massive efflux of K^+ accompanied with water loss and lowering of turgor pressure were reported during the second repolarization phase (Kishimoto and Ohkawa 1966; Barry 1970a, b; Oda 1975, 1976; discussed in Wayne 1993). Recent study reports that the depolarization phase of Chara internode action potentials is associated with significant displacements of the cellular surface (Fillafer et al. 2018). Moreover, the membrane excitation during the action potential is invariably accompanied by purely physical phenomena of structural changes in the submembraneous cytoplasm including swelling, due to a sudden rise of gel water content, and heat production (Tasaki 1999). Importantly in this respect, the lipid bilayer has unique properties (Mouritsen and Bagatolli 2015) allowing, besides action potentials, also rapid propagation of pH changes, lipid stress, and sound pulses (Fichtl et al. 2016, 2018; Aponte-Santamaría et al. 2017). In fact, action potentials are not purely electrical phenomena but more complex thermodynamic processes encompassing mechanical, optical, thermal, and chemical features (Tasaki 1999; Heimburg and Jackson 2007; Fichtl et al. 2018; Fillafer et al. 2018). Perhaps the most dramatic and obvious structural impacts of action potentials found in the large algal cells of Chara and Nitella are the sudden cessation of cytoplasmic streaming which is based on F-actin and myosin motors (Kikuyama 2001; Shimment 2007; Shimmen and Yokota 2004; Verchot-Lubicz and Goldstein 2010; Beilby 2016; Duan and Tominaga 2018). It can be expected that similar actin-based structural changes are induced via the action potentials also in other plant cells.

5.6 Actin Cytoskeleton in Electromechanical Couplings During Action Potentials

Actin cytoskeleton is acting downstream of the electrochemical phenomena, such as the action potentials, accomplished via excitable membranes. The impacts of action potentials on the actin cytoskeleton can be due to several mechanisms. First of all, the action potentials have direct impacts on the lipid bilayer phospholipids and on proteins inserted into the plasma membranes. These include ion channels and transporters, but also transmembrane proteins directly controlling actin polymerization such as integrins and formins (Deeks et al. 2005; Baluška and Hlavačka 2005; Ziegler et al. 2008; Cheung et al. 2010; Cvrčková 2013; Park and Goda 2016; Goult et al. 2018; Lan et al. 2018; Manoli et al. 2019). In maize root apices, cells of the transition zone have very high electric and synaptic activities, the latter being driven via very active endocytosis and endocytic vesicle recycling (Baluška et al. 2005, 2009, 2010; Baluška and Mancuso 2013; Masi et al. 2009, 2015). Intriguingly, these cells also assemble absolutely unique arrangement of the actin cytoskeleton, when the end-poles (cross-walls) are associated with very dense F-actin meshworks and abundant myosin VIII driving endocytosis and endocytic vesicle recycling

(Volkmann et al. 2003; Baluška et al. 1997, 2009; Baluška and Hlavačka 2005; Baluška and Mancuso 2013; Voigt et al. 2005a, b). In neurons, a single action potential was reported to induce ultrafast endocytosis at the hippocampal synapses, at time scales of 50–100 ms. This activity-dependent electromechanical coupling is based on formin-induced actin polymerization (Watanabe et al. 2014, 2018; Brockmann and Rosenmund 2016; Delvendahl et al. 2016; Soykan et al. 2017). Similarly in cells of root apex transition zone, the highest activities of electric spikes are scored in the cells, which assemble dense F-actin meshworks via formin activities to support endocytosis and endocytic vesicle recycling (Deeks et al. 2005; Baluška and Hlavačka 2005; Cvrčková 2013). Besides formins, also plant-specific myosin VIII (Reichelt et al. 1999; Volkmann et al. 2003) is associated abundantly with these F-actin enriched domains (Baluška and Hlavačka 2005), and myosin VIII is known to support endocytosis (Volkmann et al. 2003; Golomb et al. 2008; Sattarzadeh et al. 2008). Myosin VIII might be relevant for the spreading of the action potentials in plants. This plant-specific actin-based endocytic motor is also involved in the actin–myosin-based gating of plant plasmodesmata (Baluška et al. 2001a, b; Volkmann et al. 2003; Radford and White 1998, 2011; Amari et al. 2014). Plasmodesmata, similarly to gap junctions in animals and humans, couple adjacent plant cells electrochemically (Spanswick and Costerton 1967; Spanswick 1972). This feature is important not only for passing of the action potentials from cell to cell but also for synchronization of neurons in animal brains (Bennett and Zukin 2004) and root apex cells in the transition zone (Baluška and Mancuso 2013; Baluška et al. 2010).

The coupling between the upstream electrochemical events and the downstream structural processes is accomplished, among other still not identified players, also via voltage-sensitive proteins and structural sterol-based lipids organized in the form of highly structured lipid rafts (Lingwood and Simons 2010; Zhao et al. 2015). Voltage-sensitive proteins include, besides diverse ion channels, inositol-lipid phosphatases which change their activity via electrically driven conformational rearrangements (Okamura et al. 2009, 2018; Okamura and Dixon 2011; Rosasco et al. 2015). After membrane depolarization, motions of the voltage sensor domains activate PtdIns $(3,4,5)P_3$ and PtdIns $(4,5)P_2$ phosphatase activities which generate relevant phospholipids controlling dynamics/assembly of the actin cytoskeleton (Senju and Lappalainen 2019; for plant cells see Braun et al. 1999; Pleskot et al. 2013, 2014). Importantly, these actin cytoskeleton organizing phospholipids are enriched within lipid rafts acting as signaling platforms at the plasma membrane (Chichili and Rodgers 2007, 2009; Furt et al. 2010; Lingwood and Simons 2010; Klappe et al. 2013; Byrum and Rodgers 2015), and the actin cytoskeleton attached to the plasma membrane induces formation of lipid rafts (Chichili and Rodgers 2007; Dinic et al. 2013). Curiously enough, assembly and ordering of these lipid rafts are sensitive to electrochemical signaling within the plasma membrane (Lenne et al. 2006; Pristerà and Okuse 2011; Pristerà et al. 2012; Herman et al. 2015; Malinsky et al. 2016). In cells of root apices, the highest lipid order was scored in synchronized cells of the root apex transition zone (Baluška and Mancuso 2013; Zhao et al. 2015), which show the highest activities of electric spikes (Masi et al. 2009, 2015) and of the synaptic-like endocytic vesicle recycling (Baluška et al. 2005, 2009, 2010; Baluška and Mancuso 2013). The actin cytoskeleton, powered with plant-specific myosins VIII and XI, is the main player behind this active endocytic vesicle recycling which is tightly coupled with the polar cell-to-cell transport of the auxin underlying root apex tropisms (Baluška et al. 2005, 2010). These include the negative root phototropism which is based on the blue light receptor phot1 interacting with auxin transporter PIN2 (Wan et al. 2008, 2012). Our recent study revealed that the blue light-activated phot1 receptor relocalizes into the lipid rafts (Xue et al. 2018) that control the endocytic vesicle recycling in the cells of *Arabidopsis* root apex transition zone (Li et al. 2012b; Zhao et al. 2015).

5.7 Action Potentials Control Several Aspects of Plant Life and Plant Organ Movements

Although action potentials are not in focus of current interests of plant sciences, this is rather a pity, as it is obvious that action potentials, serving as rapid long-distance signaling mechanism, control not only intracellular processes related to motilities, pH, and calcium homeostasis but also almost all physiological processes such as the phloem transport and unloading, stress tolerance, wound responses, regeneration, pollination, gene expression, protein synthesis, respiration, and photosynthesis (Pickard 1973; Davies 1987; Dziubińska et al. 1989; Fromm 1991; Wildon et al. 1992; Fromm and Eschrich 1993; Fromm and Bauer 1994; Bulychev et al. 2004; Fromm and Lautner 2007; Fisahn et al. 2004; Felle and Zimmermann 2007; Koziolek et al. 2004; Hlaváčková et al. 2006; Karpiński and Szechyńska-Hebda 2010; Szechyńska-Hebda et al. 2010, 2017; Pavlovič et al. 2011; Pavlovič and Mancuso 2011; Bulychev and Komarova 2014; Hedrich et al. 2016; Białasek et al. 2017; Choi et al. 2016, 2017; Gilroy et al. 2016; Sukhov 2016; Sukhov et al. 2019). The current lack of interest in this respect is rather surprising. It is well documented by the fact that there is no mention of action potentials at all in some of the standard textbooks of plant physiology (e.g., Taiz and Zeiger 2010). The most conspicuous role of the plant action potentials in the control and animation of movements in plant organs is known for more than 50 years now (Sibaoka 1969, 1991; Böhm et al. 2016; Pavlovič et al. 2017; Scherzer et al. 2017; Hedrich and Neher 2018; Volkov 2019; Volkov et al. 2009, 2013). But even more important, and also largely ignored, is the importance of plant action potentials in the control of cell metabolism and plant physiology as discussed above.

5.8 Anesthetics Block Action Potentials: Impacts on Actin Cytoskeleton?

Anesthetics immobilize not only animals and humans but also plants, algae, and protists. In fact, all living organisms are sensitive to anesthetics (Baluška et al. 2016), and Claude Bernard used this sensitiveness to anesthetics as the basic feature of life

(Grémiaux et al. 2014). Recently, we have reported that anesthetics, including ether, xenon, and lidocaine, induce immobility of the plant organs. *Mimosa* leaves, pea tendrils, and leaf traps of Venus flytrap were all immobilized via blocking of the action potentials and actin-based endocytic vesicle recycling (Yokawa et al. 2018, 2019). Interestingly in this respect, the actin cytoskeleton accomplishes profound changes during seismonastic *Mimosa* leaf movements (Kanzawa et al. 2006). Moreover, the specific drugs depolymerizing or over-stabilizing F-actin inhibit the *Mimosa* leaf movements (Yao et al. 2008), and F-actin is fragmented by the electrical stimulation of the *Mimosa* leaves (Yao et al. 2008).

The *Mimosa* leaf system suggests that, similarly to the brain neurons and the large algal cells of Characeae, the electrochemical action potentials induce calcium transients which affect the actin cytoskeleton (Kanzawa et al. 2006; Yao et al. 2008). As a result, cytoplasmic streaming, motilities, and vesicle trafficking are all inhibited. We can expect similar scenarios with the other plant movements animated by the action potentials such as the pea tendrils or the leaf traps of *Dionaea muscipula* and *Drosera capensis* plants (Yokawa et al. 2018). Importantly, several older studies reported that anesthetics inhibit cytoplasmic streaming and other intracellular motilities (Ewart 1902; Osterhout 1952; Aasheim et al. 1974; Lavoie et al. 1989; García-Sierra and Frixione 1993), all known to be dependent on the actin cytoskeleton and myosin motors (Reichelt et al. 1999; Volkmann et al. 2003; Ueda et al. 2015; Haraguchi et al. 2018; Ryan and Nebenführ 2018). Moreover, the anesthetics target lipid rafts which are, as discussed above, closely associated with the membrane-associated actin cytoskeleton (Morrow and Parton 2005; Koshino and Takakuwa 2009; Bandeiras et al. 2013; Weinrich and Worcester 2013). In future, it will be important to study the direct impacts of anesthetics on the actin cytoskeleton in vivo using transgenic GFP lines of *Arabidopsis* seedlings (Voigt et al. 2005a), when especially their growing roots and root hairs, showing nice patterns of cytoplasmic streaming (Voigt et al. 2005b), represent an ideal model object in this respect.

5.9 Conclusions

Ever since Luigi Galvani and Alexander von Humboldt discovered the bioelectric basis of animals and plants (Galvani 1791; Humboldt 1797; Bresadola 1998; Finger et al. 2013), the field of electrophysiology represents the basic pillar of our attempts to understand life. Claude Bernard added the sensitivity to anesthetics as another life unifying principle (Bernard 1878; Grémiaux et al. 2014). Further two scientific heroes important in supporting the unity of life view are Julius von Sachs and Jagadish Chandra Bose (Bose 1913, 1926; Shepherd 2005; Kutschera 2015; Kutschera and Baluška 2015; Kutschera and Niklas 2018). Here we have reviewed several connections between the actin cytoskeleton and inherent bioelectricity of eukaryotic membranes. It is obvious that this tremendously important field is still under-investigated, and future studies will be needed to illuminate the elusive electrochemical nature of cellular life.

Acknowledgments   We thank Mary Beilby (School of Physics, The University of NSW, Sydney, Australia) for the critical reading and commenting of our chapter.

References

Aasheim G, Fink BR, Middaugh M (1974) Inhibition of rapid axoplasmic transport by procaine hydrochloride. Anesthesiology 41:549–553

Amari K, Di Donato M, Dolja VV, Heinlein M (2014) Myosins VIII and XI play distinct roles in reproduction and transport of tobacco mosaic virus. PLoS Pathog 10:e1004448

Andrews NW, Almeida PE, Corrotte M (2014) Damage control: cellular mechanisms of plasma membrane repair. Trends Cell Biol 24:734–742

Aponte-Santamaría C, Brunken J, Gräter F (2017) Stress propagation through biological lipid bilayers *in silico*. J Am Chem Soc 139:13588–13591

Arisz SA, Testerink C, Munnik T (2009) Plant PA signaling via diacylglycerol kinase. Biochim Biophys Acta 1791:869–875

Baluška F (2010) Recent surprising similarities between plant cells and neurons. Plant Signal Behav 5:87–89

Baluška F, Hlavačka A (2005) Plant formins come of age: something special about cross-walls. New Phytol 168:499–503

Baluška F, Mancuso S (2013) Root apex transition zone as oscillatory zone. Front Plant Sci 4:354

Baluška F, Wan Y-L (2012) Physical control over endocytosis. In: Šamaj J (ed) Endocytosis in plants. Springer, Berlin, pp 123–149

Baluška F, Vitha S, Barlow PW, Volkmann D (1997) Rearrangements of F-actin arrays in growing cells of intact maize root apex tissues: a major developmental switch occurs in the postmitotic transition region. Eur J Cell Biol 72:113–121

Baluška F, Šamaj J, Napier R, Volkmann D (1999) Maize calreticulin localizes preferentially to plasmodesmata in root apex. Plant J 19:481–488

Baluška F, Salaj J, Mathur J, Braun M, Jasper F, Šamaj J, Chua NH, Barlow PW, Volkmann D (2000) Root hair formation: F-actin-dependent tip growth is initiated by local assembly of profilin-supported F-actin meshworks accumulated within expansin-enriched bulges. Dev Biol 227:618–632

Baluška F, Cvrčková F, Kendrick-Jones J, Volkmann D (2001a) Sink plasmodesmata as gateways for phloem unloading. Myosin VIII and calreticulin as molecular determinants of sink strength? Plant Physiol 126:39–46

Baluška F, Jásik J, Edelmann HG, Salajová T, Volkmann D (2001b) Latrunculin B-induced plant dwarfism: plant cell elongation is F-actin-dependent. Dev Biol 231:113–124

Baluška F, Šamaj J, Hlavačka A, Kendrick-Jones J, Volkmann D (2004) Actin-dependent fluid-phase endocytosis in inner cortex cells of maize root apices. J Exp Bot 55:463–473

Baluška F, Volkmann D, Menzel D (2005) Plant synapses: actin-based domains for cell-to-cell communication. Trends Plant Sci 10:106–111

Baluška F, Barlow PW, Volkmann D, Mancuso S (2006) Gravity related paradoxes in plants: plant neurobiology provides the means for their resolution. In: Witzany G (ed) Biosemiotics in transdisciplinary context, Proceedings of the gathering in biosemiotics 6, Salzburg. Umweb, Helsinki

Baluška F, Schlicht M, Volkmann D, Mancuso S (2008) Vesicular secretion of auxin: evidences and implications. Plant Signal Behav 3:254–256

Baluška F, Schlicht M, Wan Y-L, Burbach C, Volkmann D (2009) Intracellular domains and polarity in root apices: from synaptic domains to plant neurobiology. Nova Acta Leopold 96:103–122

Baluška F, Mancuso S, Volkmann D, Barlow PW (2010) Root apex transition zone: a signalling-response nexus in the root. Trends Plant Sci 15:402–408

Baluška F, Yokawa K, Mancuso S, Baverstock K (2016) Understanding of anesthesia – why consciousness is essential for life and not based on genes. Commun Integr Biol 9:e1238118

Baluška F, Strnad M, Mancuso S (2018) Substantial evidence for auxin secretory vesicles. Plant Physiol 176:2586–2587

Bandeiras C, Serro AP, Luzyanin K, Fernandes A, Saramago B (2013) Anesthetics interacting with lipid rafts. Eur J Pharm Sci 48:153–165

Barry PH (1970a) Volume flows and pressure changes during an action potential in cells of *Chara australis*. I. Experimental results. J Membr Biol 3:313–334

Barry PH (1970b) Volume flows and pressure changes during an action potential in cells of *Chara australis*. II. Theoretical considerations. J Membr Biol 3:335–371

Bean BP (2007) The action potential in mammalian central neurons. Nat Rev Neurosci 8:451–465

Beilby MJ (1984) Calcium and plant action potentials. Plant Cell Environ 7:415–421

Beilby M (2007) Action potential in charophytes. Int Rev Cytol 257:43–82

Beilby M (2016) Multi-scale characean experimental system: from electrophysiology of membrane transporters to cell-to-cell connectivity, cytoplasmic streaming and auxin metabolism. Front Plant Sci 7:1052

Beilby MJ, Al Khazaaly S (2016) Re-modeling Chara action potential: I. From Thiel model of Ca^{2+} transient to action potential form. AIMS Biophys 3:431–449

Beilby MJ, Shepherd VA (1996) Turgor regulation in *Lamprothamnium papulosum*.1. I/V analysis and pharmacological dissection of the hypotonic effect. Plant Cell Environ 19:837–847

Bennett MVL, Zukin RS (2004) Electrical coupling and neuronal synchronization in the mammalian brain. Neuron 41:495–511

Berghöfer T, Eing C, Flickinger B, Hohenberger P, Wegner LH, Frey W, Nick P (2009) Nanosecond electric pulses trigger actin responses in plant cells. Biochem Biophys Res Commun 387:590–595

Bernard C (1878) Leçonssur les phénomènes de la vie communs aux animauxet aux végétaux. Lectures on phenomena of life common to animals and plants. Ballliere, Paris

Bezanilla F (2006) The action potential: from voltage-gated conductances to molecular structures. Biol Res 39:425–435

Bezanilla F (2008) Ion channels: from conductance to structure. Neuron 60:456–468

Białasek M, Górecka M, Mittler R, Karpiński S (2017) Evidence for the involvement of electrical, calcium and ROS signaling in the systemic regulation of non-photochemical quenching and photosynthesis. Plant Cell Physiol 58:207–215

Blinks LR, Harris ES, Osterhout WJV (1929) Studies on stimulation in Nitella. Proc Soc Exp Biol Med 26:836–838

Böhm J, Scherzer S, Krol E, Kreuzer I, von Meyer K, Lorey C, Mueller TD, Shabala L, Monte I, Solano R, Al-Rasheid KA, Rennenberg H, Shabala S, Neher E, Hedrich R (2016) The Venus flytrap *Dionaea muscipula* counts prey-induced action potentials to induce sodium uptake. Curr Biol 26:286–295

Bose JC (1913) Researches on irritability of plants. Longmans, Green, London

Bose JC (1926) The nervous mechanisms of plants. Longmans, Green, London

Boye TL, Nylandsted J (2016) Annexins in plasma membrane repair. Biol Chem 397:961–969

Boye TL, Maeda K, Pezeshkian W, Sønder SL, Haeger SC, Gerke V, Simonsen AC, Nylandsted J (2017) Annexin A4 and A6 induce membrane curvature and constriction during cell membrane repair. Nat Commun 8:1623

Braun M, Baluška F, von Witsch M, Menzel D (1999) Redistribution of actin, profilin and phosphatidylinositol-4,5-bisphosphate in growing and maturing root hairs. Planta 209:435–443

Bresadola M (1998) Medicine and science in the life of Luigi Galvani (1737–1798). Brain Res Bull 46:367–380

Brockmann MM, Rosenmund C (2016) Catching up with ultrafast endocytosis. Neuron 90:423–424

Brunet T, Arendt D (2016) From damage response to action potentials: early evolution of neural and contractile modules in stem eukaryotes. Philos Trans R Soc Lond Ser B Biol Sci 371:20150043

Bulychev AA, Komarova AV (2014) Long-distance signal transmission and regulation of photosynthesis in characean cells. Biochem Mosc 79:273–281

Bulychev AA, Kamzolkina NA, Luengviriya J, Rubin AB, Müller SC (2004) Effect of a single excitation stimulus on photosynthetic activity and light-dependent pH banding in Chara cells. J Membr Biol 202:11–19

Byrum JN, Rodgers W (2015) Membrane-cytoskeleton interactions in cholesterol-dependent domain formation. Essays Biochem 57:177–187

Cheung AY, Niroomand S, Zou Y, Wu HM (2010) A transmembrane formin nucleates subapical actin assembly and controls tip-focused growth in pollen tubes. Proc Natl Acad Sci USA 107:16390–16395

Chichili GR, Rodgers W (2007) Clustering of membrane raft proteins by the actin cytoskeleton. J Biol Chem 282:36682–36691

Chichili GR, Rodgers W (2009) Cytoskeleton-membrane interactions in membrane raft structure. Cell Mol Life Sci 66:2319–2328

Choi WG, Hilleary R, Swanson SJ, Kim SH, Gilroy S (2016) Rapid, long-distance electrical and calcium signaling in plants. Annu Rev Plant Biol 67:287–307

Choi WG, Miller G, Wallace I, Harper J, Mittler R, Gilroy S (2017) Orchestrating rapid long-distance signaling in plants with Ca^{2+}, ROS and electrical signals. Plant J 90:698–707

Clark GB, Morgan RO, Fernandez MP, Roux SJ (2012) Evolutionary adaptation of plant annexins has diversified their molecular structures, interactions and functional roles. New Phytol 196:695–712

Clayton L, Lloyd CW (1985) Actin organization during the cell cycle in meristematic plant cells. Actin is present in the cytokinetic phragmoplast. Exp Cell Res 156:231–238

Cohen LB (1973) Changes in neuron structure during action potential propagation and synaptic transmission. Physiol Rev 53:373–418

Cole KS, Curtis HJ (1938) Electric impedance of Nitella during activity. J Gen Physiol 22:37–64

Cole KS, Curtis HJ (1939) Electric impedance of the squid giant axon during activity. J Gen Physiol 22:649–670

Craxton M (2007) Evolutionary genomics of plant genes encoding N-terminal-TM-C2 domain proteins and the similar FAM62 genes and synaptotagmin genes of metazoans. BMC Genomics 8:259

Cvrčková F (2013) Formins and membranes: anchoring cortical actin to the cell wall and beyond. Front Plant Sci 4:436

Davies E (1987) Action potentials as multifunctional signals in plants: a unifying hypothesis to explain apparently disparate wound responses. Plant Cell Environ 10:623–631

Davies JM (2014) Annexin-mediated calcium signalling in plants. Plan Theory 3:128–140

Deeks MJ, Cvrčková F, Machesky LM, Mikitová V, Ketelaar T, Zársky V, Davies B, Hussey PJ (2005) Arabidopsis group Ie formins localize to specific cell membrane domains, interact with actin-binding proteins and cause defects in cell expansion upon aberrant expression. New Phytol 168:529–540

Delvendahl I, Vyleta NP, von Gersdorff H, Hallermann S (2016) Fast, temperature-sensitive and clathrin-independent endocytosis at central synapses. Neuron 90:492–498

Dhonukshe P, Grigoriev I, Fischer R, Tominaga M, Robinson DG, Hasek J, Paciorek T, Petrásek J, Seifertová D, Tejos R, Meisel LA, Zazímalová E, Gadella TW Jr, Stierhof YD, Ueda T, Oiwa K, Akhmanova A, Brock R, Spang A, Friml J (2008) Auxin transport inhibitors impair vesicle motility and actin cytoskeleton dynamics in diverse eukaryotes. Proc Natl Acad Sci USA 105:4489–4494

Diao M, Ren S, Wang Q, Qian L, Shen J, Liu Y, Huang S (2018) Arabidopsis formin 2 regulates cell-to-cell trafficking by capping and stabilizing actin filaments at plasmodesmata. elife 7: e36316

Dinic J, Ashrafzadeh P, Parmryd I (2013) Actin filaments attachment at the plasma membrane in live cells cause the formation of ordered lipid domains. Biochim Biophys Acta 1828:1102–1111

Dong W, Lv H, Xia G, Wang M (2012) Does diacylglycerol serve as a signaling molecule in plants? Plant Signal Behav 7:472–475

Duan Z, Tominaga M (2018) Actin-myosin XI: an intracellular control network in plants. Biochem Biophys Res Commun 506:403–408

Dünser K, Gupta S, Herger A, Feraru MI, Ringli C, Kleine-Vehn J (2019) Extracellular matrix sensing by FERONIA and leucine-rich repeat extensins controls vacuolar expansion during cellular elongation in *Arabidopsis thaliana*. EMBO J 38:e100353

Dziubińska H, Trębacz K, Zawadzki T (1989) The effect of excitation on the rate of respiration in the liverwort *Conocephalum conicum*. Physiol Plant 75:417–423

Ewart AJ (1902) On the physics and physiology of the protoplasmic streaming in plants. Proc R Soc Lond 69:466–470

Ewart AJ (1903) On the physics and physiology of protoplasmic streaming in plants. Clarendon Press, Oxford

Felle HH, Zimmermann MR (2007) Systemic signalling in barley through action potentials. Planta 226:203–214

Fichtl B, Shrivastava S, Schneider MF (2016) Protons at the speed of sound: predicting specific biological signaling from physics. Sci Rep 6:22874

Fichtl B, Silman I, Schneider MF (2018) On the physical basis of biological signaling by interface pulses. Langmuir 34:4914–4919

Fillafer C, Mussel M, Muchowski J, Schneider MF (2018) Cell surface deformation during an action potential. Biophys J 114:410–418

Findlay GP, Hope AB (1964) Ionic relations of cells of *Chara australis*: VII. The separate electrical characteristics of the plasmalemma and the tonoplast. Aust J Biol Sci 17:62–77

Finger S, Piccolino M, Stahnisch FW (2013) Alexander von Humboldt: Galvanism, animal electricity, and self-experimentation. I. Formative years, naturphilosophie, and Galvanism. J Hist Neurosci Basic Clin Perspect 22:225–260

Fink BR, Kish SJ (1976) Reversible inhibition of rapid axonal transport in vivo by lidocaine hydrochloride. Anesthesiology 44:139–146

Fisahn J, Herde O, Willmitzer L, Peña-Cortés H (2004) Analysis of the transient increase in cytosolic Ca^{2+} during the action potential of higher plants with high temporal resolution: requirement of Ca^{2+} transients for induction of jasmonic acid biosynthesis and PINII gene expression. Plant Cell Physiol 45:456–459

Foissner I, Wasteneys GO (2012) The characean internodal cell as a model system for studying wound healing. J Microsc 247:10–22

Fromm J (1991) Control of phloem unloading by action potentials in Mimosa. Physiol Plant 83:529–533

Fromm J, Bauer T (1994) Action potentials in maize sieve tubes change phloem translocation. J Exp Bot 45:463–469

Fromm J, Eschrich W (1993) Electric signals released from roots of willow (*Salix viminalis* L.) change transpiration and photosynthesis. J Plant Physiol 141:673–680

Fromm J, Lautner S (2007) Electrical signals and their physiological significance in plants. Plant Cell Environ 30:249–257

Furt F, König S, Bessoule J-J, Sargueil F, Zallot R, Stanislas T, Noirot E, Lherminier J, Simon-Plas F, Heilmann I, Mongrand S (2010) Polyphosphoinositides are enriched in plant membrane rafts and form microdomains in the plasma membrane. Plant Physiol 152:2173–2218

Galvani L (1791, 1953) Commentary on the effect of electricity on muscular motion (translation of *De viribus electricitatis in motu musculari commentarius* by Green RM). Norwalk: Burndy Library

Gao XQ, Wang XL, Ren F, Chen J, Wang XC (2009) Dynamics of vacuoles and actin filaments in guard cells and their roles in stomatal movement. Plant Cell Environ 32:1108–1116

García-Sierra F, Frixione E (1993) Lidocaine, a local anesthetic, reversibly inhibits cytoplasmic streaming in *Vallisneria* mesophyll cells. Protoplasma 175:153–160

Gilroy S, Białasek M, Suzuki N, Gorecka M, Devireddy A, Karpinski S, Mittler R (2016) ROS, calcium and electric signals: key mediators of rapid systemic signaling in plants. Plant Physiol 171:1606–1615

Goldsworthy A (1983) The evolution of plant action potentials. J Theor Biol 103:645–648

Golomb L, Abu-Abied M, Belausov E, Sadot E (2008) Different subcellular localizations and functions of Arabidopsis myosin VIII. BMC Plant Biol 8:3

Goult BT, Yan J, Schwartz MA (2018) Talin as a mechanosensitive signaling hub. J Cell Biol 217:3776–3784

Grémiaux A, Yokawa K, Mancuso S, Baluška F (2014) Plant anesthesia supports similarities between animals and plants: Claude Bernard's forgotten studies. Plant Signal Behav 9:e27886

Haraguchi T, Ito K, Duan Z, Rula S, Takahashi K, Shibuya Y, Hagino N, Miyatake Y, Nakano A, Tominaga M (2018) Functional diversity of class XI myosins in *Arabidopsis thaliana*. Plant Cell Physiol 59:2268–2277

Hedrich R (2012) Ion channels in plants. Physiol Rev 92:1777–1811

Hedrich R, Neher E (2018) Venus flytrap: how an excitable, carnivorous plant works. Trends Plant Sci 23:220–234

Hedrich R, Salvador-Recatalà V, Dreyer I (2016) Electrical wiring and long-distance plant communication. Trends Plant Sci 21:376–387

Heidecker M, Wegner LH, Binder K, Zimmermann U (2003) Turgor pressure changes trigger characteristic changes in the electrical conductance of the tonoplast and the plasmalemma of the marine alga *Valonia utricularis*. Plant Cell Environ 26:1035–1051

Heimburg T, Jackson AD (2007) On the action potential as a propagating density pulse and the role of anesthetics. Biophys Rev Lett 2:57–78

Herman P, Vecer J, Opekarova M, Vesela P, Jancikova I, Zahumensky J, Malinsky J (2015) Depolarization affects the lateral microdomain structure of yeast plasma membrane. FEBS J 282:419–434

Hill SE (1941) The relation between protoplasmic streaming and action potential in Nitella and Chara. Biol Bull 81:296–303

Hirsch RE, Lewis BD, Spalding EP, Sussman MR (1989) A role for the AKT1 potassium channel in plant nutrition. Science 280:918–921

Hlaváčková V, Krchňák P, Naus J, Novák O, Spundová M, Strnad M (2006) Electrical and chemical signals involved in short-term systemic photosynthetic responses of tobacco plants to local burning. Planta 225:235–244

Hörmann G (1898) Studien über die Protoplasmaströmung bei den Characeen. Gustav Fischer Verlag, Jena

Humboldt A (1797) Versuche über die gereizte Muskel- und Nervenfaser nebst Vermuthungen über den chemischen Process des Lebens in der Thier- und Pflanzenwelt. Posen, Decker und Compagnie, Berlin

Iwasa K, Tasaki I, Gibbons RC (1980) Swelling of nerve fibers associated with action potential. Science 210:338–339

Kanzawa N, Hoshino Y, Chiba M, Hoshino D, Kobayashi H, Kamasawa N, Kishi Y, Osumi M, Sameshima M, Tsuchiya T (2006) Change in the actin cytoskeleton during seismonastic movement of *Mimosa pudica*. Plant Cell Physiol 47:531–539

Karpiński S, Szechyńska-Hebda M (2010) Secret life of plants: from memory to intelligence. Plant Signal Behav 5:1391

Kersey YM, Hepler PK, Palevitz BA, Wessells NK (1976) Polarity of actin filaments in Characean algae. Proc Natl Acad Sci USA 73:165–167

Keynes RD (1958) The nerve impulse and the squid. Sci Am 199:83–90

Kikuyama M (1986) Tonoplast action potential of Characeae. Plant Cell Physiol 27:1461–1468

Kikuyama M (2001) Role of Ca^{2+} in membrane excitation and cell motility in Characean cells as a model system. Int Rev Cytol 201:85–114

Kikuyama M, Shimmen M (1997) Role of Ca^{2+} on triggering tonoplast action potential in intact *Nitella flexilis*. Plant Cell Physiol 38:941–944

Kikuyama M, Tazawa M (1976) Tonoplast action potential in Nitella in relation to vacuolar chloride concentration. J Membr Biol 29:95–110

Kishimoto U, Akabori H (1959) Protoplasmic streaming of an internodal cell of *Nitella flexilis*; its correlation with electric stimulus. J Gen Physiol 42:1167–1183

Kishimoto U, Ohkawa T (1966) Shortening of Nitella internode during excitation. Plant Cell Physiol 7:493–497

Kisnieriene V, Lapeikaite I, Pupkis V, Beilby MJ (2019) Modeling the action potential in characeae *Nitellopsis obtusa*: effect of saline stress. Front Plant Sci 10:82

Klappe K, Hummel I, Kok JW (2013) Separation of actin-dependent and actin-independent lipid rafts. Anal Biochem 438:133–135

Koshino I, Takakuwa Y (2009) Disruption of lipid rafts by lidocaine inhibits erythrocyte invasion by *Plasmodium falciparum*. Exp Parasitol 123:381–383

Koziolek C, Grams TEE, Schreiber U, Matyssek R, Fromm J (2004) Transient knockout of photosynthesis mediated by electrical signals. New Phytol 161:715–722

Kutschera U (2015) Comment: 150 years of an integrative plant physiology. Nat Plants 1:1–3

Kutschera U, Baluška F (2015) Julius Sachs (1832–1897) and the unity of life. Plant Signal Behav 10:e1079679

Kutschera U, Niklas KJ (2018) Julius Sachs (1868): the father of plant physiology. Am J Bot 105:656–666

Lan Y, Liu X, Fu Y, Huang S (2018) Arabidopsis class I formins control membrane-originated actin polymerization at pollen tube tips. PLoS Genet 14:e1007789

Lavoie P-A, Khazen T, Filion PR (1989) Mechanisms of inhibition of fast axonal transport by local anesthetics. Neuropharmacology 28:175–181

Lenne PF, Wawrezinieck L, Conchonaud F, Wurtz O, Boned A, Guo XJ, Rigneault H, He HT, Marguet D (2006) Dynamic molecular confinement in the plasma membrane by microdomains and the cytoskeleton meshwork. EMBO J 25:3245–3256

Li J, Staiger CJ (2018) Understanding cytoskeletal dynamics during the plant immune response. Annu Rev Phytopathol 56:513–533

Li J, Pleskot R, Henty-Ridilla JL, Blanchoin L, Potocký M, Staiger CJ (2012a) Arabidopsis capping protein senses cellular phosphatidic acid levels and transduces these into changes in actin cytoskeleton dynamics. Plant Signal Behav 7:1727–1730

Li R, Liu P, Wan Y, Chen T, Wang Q, Mettbach U, Baluška F, Samaj J, Fang X, Lucas WJ, Lin J (2012b) A membrane microdomain-associated protein, Arabidopsis Flot1, is involved in a clathrin-independent endocytic pathway and is required for seedling development. Plant Cell 24:2105–2122

Li LJ, Ren F, Gao XQ, Wei PC, Wang XC (2013) The reorganization of actin filaments is required for vacuolar fusion of guard cells during stomatal opening in Arabidopsis. Plant Cell Environ 36:484–497

Li J, Blanchoin L, Staiger CJ (2015) Signaling to actin stochastic dynamics. Annu Rev Plant Biol 66:415–440

Lingwood D, Simons K (2010) Lipid rafts as a membrane-organizing principle. Science 327:46–50

Malinsky J, Tanner W, Opekarova M (2016) Transmembrane voltage: potential to induce lateral microdomains. Biochim Biophys Acta 1861:806–811

Mancuso S, Marras AM, Magnus V, Baluška F (2005) Noninvasive and continuous recordings of auxin fluxes in intact root apex with a carbon nanotube-modified and self-referencing micro-electrode. Anal Biochem 341:344–351

Mancuso S, Marras AM, Mugnai S, Schlicht M, Žárský V, Li G, Song L, Xue HW, Baluška F (2007) Phospholipase dzeta2 drives vesicular secretion of auxin for its polar cell-cell transport in the transition zone of the root apex. Plant Signal Behav 2:240–244

Manoli S, Coppola S, Duranti C, Lulli M, Magni L, Kuppalu N, Nielsen N, Schmidt T, Schwab A, Becchetti A, Arcangeli A (2019) The activity of Kv 11.1 potassium channel modulates F-actin organization during cell migration of pancreatic ductal adenocarcinoma cells. Cancers (Basel) 11(2):e135

Masi E, Ciszak M, Stefano G, Renna L, Azzarello E, Pandolfi C, Mugnai S, Baluška F, Arecchi FT, Mancuso S (2009) Spatiotemporal dynamics of the electrical network activity in the root apex. Proc Natl Acad Sci USA 106:4048–4053

Masi E, Ciszak M, Comparini D, Monetti E, Pandolfi C, Azzarello E, Mugnai S, Baluška F, Mancuso S (2015) The electrical network of maize root apex is gravity dependent. Sci Rep 5:7730

Menzel D (1988) How do giant plant cells cope with injury? The wound response in siphonous green algae. Protoplasma 144:73–91

Mettbach U, Strnad M, Mancuso S, Baluška F (2017) Immunogold-EM analysis reveal Brefeldin A-sensitive clusters of auxin in Arabidopsis root apex cells. Commun Integr Biol 10:e1327105

Morrow IC, Parton RG (2005) Flotillins and the PHB domain protein family: rafts, worms and anaesthetics. Traffic 6:725–740

Mouritsen OG, Bagatolli LA (2015) Lipid domains in model membranes: a brief historical perspective. Essays Biochem 57:1–19

Němec B (1901) Die Reizleitung und die Reizleitenden Strukturen bei den Pflanzen. Verlag von Gustaf Fischer, Jena

Oda K (1975) Recording of the potassium efflux during a single action potential in Chara corallina. Plant Cell Physiol 16:525–528

Oda K (1976) Simultaneous recording of potassium and chloride effluxes during an action potential in Chara corallina. Plant Cell Physiol 17:1085–1088

Okamura Y, Dixon JE (2011) Voltage-sensing phosphatase: its molecular relationship with PTEN. Physiology 26:6–13

Okamura Y, Murata Y, Iwasaki H (2009) Voltage-sensing phosphatase: actions and potentials. J Physiol 587:513–520

Okamura Y, Kawanabe A, Kawai T (2018) Voltage-sensing phosphatases: biophysics, physiology, and molecular engineering. Physiol Rev 98:2097–2131

Osterhout WJV (1936) Electrical phenomena in large plant cells. Physiol Rev 16:216–237

Osterhout WJV (1952) Some aspects of protoplasmic motion. J Gen Physiol 35:519–527

Palevitz BA, Hepler PK (1975) Identification of actin in situ at the ectoplasm-endoplasm interface of Nitella. Microfilament-chloroplast association. J Cell Biol 65:29–38

Palevitz BA, Ash JF, Hepler PK (1974) Actin in the green alga, Nitella. Proc Natl Acad Sci USA 71:363–366

Park YK, Goda Y (2016) Integrins in synapse regulation. Nat Rev Neurosci 17:745–756

Pavlovič A, Mancuso S (2011) Electrical signaling and photosynthesis: can they co-exist together? Plant Signal Behav 6:840–842

Pavlovič A, Slováková L, Pandolfi C, Mancuso S (2011) On the mechanism underlying photosynthetic limitation upon trigger hair irritation in the carnivorous plant Venus flytrap (Dionaea muscipula Ellis). J Exp Bot 62:1991–2000

Pavlovič A, Jakšová J, Novák O (2017) Triggering a false alarm: wounding mimics prey capture in the carnivorous Venus flytrap (Dionaea muscipula). New Phytol 216:927–938

Pedersen CN, Axelsen KB, Harper JF, Palmgren MG (2012) Evolution of plant P-type ATPases. Front Plant Sci 3:31

Pérez-Sancho J, Vanneste S, Lee E, McFarlane HE, Esteban Del Valle A, Valpuesta V, Friml J, Botella MA, Rosado A (2015) The Arabidopsis synaptotagmin1 is enriched in endoplasmic reticulum-plasma membrane contact sites and confers cellular resistance to mechanical stresses. Plant Physiol 168:132–143

Pickard BG (1973) Action potentials in higher plants. Bot Rev 39:172–201

Pleskot R, Pejchar P, Žárský V, Staiger CJ, Potocký M (2012) Structural insights into the inhibition of actin-capping protein by interactions with phosphatidic acid and phosphatidylinositol (4,5)-bisphosphate. PLoS Comput Biol 8(11):e1002765

Pleskot R, Li J, Žárský V, Potocký M, Staiger CJ (2013) Regulation of cytoskeletal dynamics by phospholipase D and phosphatidic acid. Trends Plant Sci 18:496–504

Pleskot R, Pejchar P, Staiger CJ, Potocký M (2014) When fat is not bad: the regulation of actin dynamics by phospholipid signaling molecules. Front Plant Sci 5:5

Pristerà A, Okuse K (2011) Building excitable membranes: lipid rafts and multiple controls on trafficking of electrogenic molecules. Neuroscientist 18:70–81

Pristerà A, Baker MD, Okuse K (2012) Association between tetrodotoxin resistant channels and lipid rafts regulates sensory neuron excitability. PLoS One 7:e40079

Radford JE, White RG (1998) Localization of a myosin-like protein to plasmodesmata. Plant J 14:743–750

Radford JE, White RG (2011) Inhibitors of myosin, but not actin, alter transport through *Tradescantia* plasmodesmata. Protoplasma 248:205–216

Reichelt S, Knight AE, Hodge TP, Baluška F, Šamaj J, Volkmann D, Kendrick-Jones J (1999) Characterization of the unconventional myosin VIII in plant cells and its localization at the post-cytokinetic cell wall. Plant J 19:555–567

Richards SL, Laohavisit A, Mortimer JC, Shabala L, Swarbreck SM, Shabala S, Davies JM (2014) Annexin 1 regulates the H2O2-induced calcium signature in *Arabidopsis thaliana* roots. Plant J 77:136–145

Roelfsema MR, Steinmeyer R, Staal M, Hedrich R (2001) Single guard cell recordings in intact, plants: light-induced hyperpolarization of the plasma membrane. Plant J 26:1–13

Rosasco MG, Gordon SE, Bajjalieh SM (2015) Characterization of the functional domains of a mammalian voltage-sensitive phosphatase. Biophys J 109:2480–2491

Ryan JM, Nebenführ A (2018) Update on myosin motors: molecular mechanisms and physiological functions. Plant Physiol 176:119–127

Saheki Y, De Camilli P (2017) The extended synaptotagmins. Biochim Biophys Acta Mol Cell Res 1864:1490–1493

Šamaj J, Read ND, Volkmann D, Menzel D, Baluška F (2005) The endocytic network in plants. Trends Cell Biol 15:425–433

Sattarzadeh A, Franzen R, Schmelzer E (2008) The Arabidopsis class VIII myosin ATM2 is involved in endocytosis. Cell Motil Cytoskeleton 65:457–468

Schapire AL, Voigt B, Jasik J, Rosado A, Lopez-Cobollo R, Menzel D, Salinas J, Mancuso S, Valpuesta V, Baluška F, Botella MA (2008) Arabidopsis synaptotagmin 1 is required for the maintenance of plasma membrane integrity and cell viability. Plant Cell 20:3374–3388

Schapire AL, Valpuesta V, Botella MA (2009) Plasma membrane repair in plants. Trends Plant Sci 14:645–652

Scherzer S, Shabala L, Hedrich B, Fromm J, Bauer H, Munz E, Jakob P, Al-Rascheid KAS, Kreuzer I, Becker D, Eiblmeier M, Rennenberg H, Shabala S, Bennett M, Neher E, Hedrich R (2017) Insect haptoelectrical stimulation of Venus flytrap triggers exocytosis in gland cells. Proc Natl Acad Sci USA 114:4822–4827

Scheuring D, Löfke C, Krüger F, Kittelmann M, Eisa A, Hughes L, Smith RS, Hawes C, Schumacher K, Kleine-Vehn J (2016) Actin-dependent vacuolar occupancy of the cell determines auxin-induced growth repression. Proc Natl Acad Sci USA 113:452–457

Schlicht M, Strnad M, Scanlon MJ, Mancuso S, Hochholdinger F, Palme K, Volkmann D, Menzel D, Baluška F (2006) Auxin immunolocalization implicates vesicular neurotransmitter-like mode of polar auxin transport in root apices. Plant Signal Behav 1:122–133

Seagull RW, Falconer MM, Weerdenburg CA (1987) Microfilaments: dynamic arrays in higher plant cells. J Cell Biol 104:995–1004

Senju Y, Lappalainen P (2019) Regulation of actin dynamics by PI(4,5)P2 in cell migration and endocytosis. Curr Opin Cell Biol 56:7–13

Shepherd VA (2005) From semi-conductors to the rhythms of sensitive plants: the research of J. C. Bose. Cell Mol Biol 51:607–619

Shepherd VA, Beilby MJ, Al Khazaaly SA, Shimmen T (1998) Mechano-perception in Chara cells: the influence of salinity and calcium on touch-activated receptor potentials, action potentials and ion transport. Plant Cell Environ 31:1575–1591

Shepherd VA, Beilby MJ, Bisson MA (2004) When is a cell not a cell? A theory relating coenocytic structure to the unusual electrophysiology of *Ventricaria ventricosa* (*Valonia ventricosa*). Protoplasma 223:79–91

Shimmen T, Nishikawa S (1988) Studies on the tonoplast action potential of *Nitella flexilis*. J Membr Biol 101:133–140

Shimmen T, Yokota E (2004) Cytoplasmic streaming in plants. Curr Opin Cell Biol 16:68–72

Shimment T (2007) The sliding theory of cytoplasmic streaming: fifty years of progress. J Plant Res 120:31–43

Siao W, Wang P, Voigt B, Hussey PJ, Baluška F (2016) Arabidopsis SYT1 maintains stability of cortical endoplasmic reticulum networks and VAP27-1-enriched endoplasmic reticulum-plasma membrane contact sites. J Exp Bot 67:6161–6171

Sibaoka T (1969) Physiology of rapid movements in higher plants. Annu Rev Plant Physiol 20:165–184

Sibaoka T (1991) Rapid plant movements triggered by action potentials. Bot Mag Tokyo 104:73–95

Sibaoka T, Oda K (1956) Shock stoppage of the protoplasmic streaming in relation to the action potential in Chara. Sci Rep Tohoku Univ IV Biol 22:157–166

Simonsen AC, Boye TL, Nylandsted J (2019) Annexins bend wound edges during plasma membrane repair. Curr Med Chem (in press)

Soykan T, Kaempf N, Sakaba T, Vollweiter D, Goerdeler F, Puchkov D, Kononenko NL, Haucke V (2017) Synaptic vesicle endocytosis occurs on multiple timescales and is mediated by formin-dependent actin assembly. Neuron 93:854–866

Spanswick RM (1972) Electrical coupling between cells of higher plants: a direct demonstration of intercellular communication. Planta 102:215–227

Spanswick RM, Costerton JW (1967) Plasmodesmata in *Nitella translucens*: structure and electrical resistance. J Cell Sci 2:451–464

Stahlberg R (2006) Historical overview on plant neurobiology. Plant Signal Behav 1:6–8

Staiger CJ, Baluška F, Volkmann D, Barlow PW (2000) Actin – a dynamic framework for multiple plant cell functions. Kluwer Academic, Dordrecht

Steinhardt RA, Bi G, Alderton JM (1994) Cell membrane resealing by a vesicular mechanism similar to neurotransmitter release. Science 263:390–394

Straub FB (1942) Actin. Stud Inst Med Chem Univ Szeged 2:3–15

Sukhov V (2016) Electrical signals as mechanism of photosynthesis regulation in plants. Photosynth Res 130:373–387

Sukhov V, Sukhova E, Vodeneev V (2019) Long-distance electrical signals as a link between the local action of stressors and the systemic physiological responses in higher plants. Prog Biophys Mol Biol 146:63–84

Szechyńska-Hebda M, Kruk J, Górecka M, Karpińska B, Karpiński S (2010) Evidence for light wavelength-specific photoelectrophysiological signaling and memory of excess light episodes in Arabidopsis. Plant Cell 22:2201–2218

Szechyńska-Hebda M, Lewandowska M, Karpiński S (2017) Electrical signaling, photosynthesis and systemic acquired acclimation. Front Physiol 8:684

Szent-Györgyi A (1942) The contraction of myosin threads. Stud Inst Med Chem Univ Szeged 1:17–26

Szent-Györgyi A (1943) Observations on actomyosin. Stud Inst Med Chem Univ Szeged 3:86–92

Szent-Györgyi AG (2004) The early history of the biochemistry of muscle contraction. J Gen Physiol 123:631–641

Taiz L, Zeiger E (2010) Plant physiology, 5th edn. Sinauer Associates, Sunderland

Takatsuka H, Higaki T, Umeda M (2018) Actin reorganization triggers rapid cell elongation in roots. Plant Physiol 178:1130–1141

Tang SKY, Marshall WF (2017) Self-repairing cells: how single cells heal membrane ruptures and restore lost structures. Science 356:1022–1025

Tasaki I (1999) Evidence for phase transition in nerve fibers, cells and synapses. Ferroelectrics 220:305–316

Tasaki I, Iwasa K, Gibbons RC (1980) Mechanical changes in crab nerve fibers during action potential. Jpn J Physiol 30:897–905

Tazawa M, Kishimoto U (1968) Cessation of cytoplasmic streaming of Chara internodes during action potential. Plant Cell Physiol 9:361–368

Togo T, Alderton JM, Bi GQ, Steinhardt RA (1999) The mechanism of facilitated cell membrane resealing. J Cell Sci 112:719–731

Traas JA, Doonan JH, Rawlins DJ, Shaw PJ, Watts J, Lloyd CW (1987) An actin network is present in the cytoplasm throughout the cell cycle of carrot cells and associates with the dividing nucleus. J Cell Biol 105:387–395

Ueda H, Tamura K, Hara-Nishimura I (2015) Functions of plant-specific myosin XI: from intracellular motility to plant postures. Curr Opin Plant Biol 28:30–38

Umrath K (1932) Der Erregungsvorgang bei *Nitella mucronata*. Protoplasma 17:258–300

Verchot-Lubicz J, Goldstein RE (2010) Cytoplasmic streaming enables the distribution of molecules and vesicles in large plant cells. Protoplasma 240:99–107

Vermeer JEM, van Wijk R, Goedhart J, Geldner N, Chory J, Gadella TWJ Jr, Munnik T (2017) In vivo imaging of diacylglycerol at the cytoplasmic leaflet of plant membranes. Plant Cell Physiol 58:1196–1207

Voigt B, Timmers AC, Šamaj J, Müller J, Baluška F, Menzel D (2005a) GFP-FABD2 fusion construct allows in vivo visualization of the dynamic actin cytoskeleton in all cells of Arabidopsis seedlings. Eur J Cell Biol 84:595–608

Voigt B, Timmers AC, Šamaj J, Hlavacka A, Ueda T, Preuss M, Nielsen E, Mathur J, Emans N, Stenmark H, Nakano A, Baluška F, Menzel D (2005b) Actin-based motility of endosomes is linked to the polar tip growth of root hairs. Eur J Cell Biol 84:609–621

Volkmann D, Baluška F (1999) Actin cytoskeleton in plants: from transport networks to signaling networks. Microsc Res Tech 47:135–154

Volkmann D, Mori T, Tirlapur UK, König K, Fujiwara T, Kendrick-Jones J, Baluška F (2003) Unconventional myosins of the plant-specific class VIII: endocytosis, cytokinesis, plasmodesmata/pit-fields, and cell-to-cell coupling. Cell Biol Int 27:289–291

Volkov AG (2019) Signaling in electrical networks of the Venus flytrap (*Dionaea muscipula* Ellis). Bioelectrochemistry 125:25–32

Volkov AG, Carrell H, Markin VS (2009) Biologically closed electrical circuits in Venus flytrap. Plant Physiol 149:1661–1667

Volkov AG, Vilfranc C, Murphy VA, Mitchell CM, Volkova MI, O'Neal L, Markin VS (2013) Electrotonic and action potentials in the Venus flytrap. J Plant Physiol 170:838–846

Wacke M, Thiel G (2001) Electrically triggered all-or-none Ca^{2+} liberation during action potential in the giant alga Chara. J Gen Physiol 118:11–21

Wacke M, Thiel G, Hutt MT (2003) Ca^{2+} dynamics during membrane excitation of green alga Chara: model simulations and experimental data. J Membr Biol 191:179–192

Walker DJ, Leigh RA, Miller AJ (2006) Potassium homeostasis in vacuolate plant cells. Proc Natl Acad Sci USA 93:10510–10514

Wan YL, Eisinger W, Ehrhardt D, Kubitscheck U, Baluška F, Briggs W (2008) The subcellular localization and blue-light-induced movement of phototropin 1-GFP in etiolated seedlings of *Arabidopsis thaliana*. Mol Plant 1:103–117

Wan Y, Jasik J, Wang L, Hao H, Volkmann D, Menzel D, Mancuso S, Baluška F, Lin J (2012) The signal transducer NPH3 integrates the phototropin1 photosensor with PIN2-based polar auxin transport in Arabidopsis root phototropism. Plant Cell 24:551–565

Wang P, Hussey PJ (2015) Interactions between plant endomembrane systems and the actin cytoskeleton. Front Plant Sci 6:422

Wang P, Hawes C, Hussey PJ (2017a) Plant endoplasmic reticulum-plasma membrane contact sites. Trends Plant Sci 22:289–297

Wang P, Hawkins TJ, Hussey PJ (2017b) Connecting membranes to the actin cytoskeleton. Curr Opin Plant Biol 40:71–76

Wang P, Hawes C, Richardson C, Hussey PJ (2018) Characterization of proteins localized to plant ER-PM contact sites. Methods Mol Biol 1691:23–31

Watanabe S, Trimbuch T, Camacho-Pérez M, Rost BR, Brokowski B, Söhl-Kielczynski B, Felies A, Davis MW, Rosenmund C, Jorgensen EM (2014) Clathrin regenerates synaptic vesicles from endosomes. Nature 515:228–233

Watanabe S, Mamer LE, Raychaudhuri S, Luvsanjav D, Eisen J, Trimbuch T, Söhl-Kielczynski B, Fenske P, Milosevic I, Rosenmund C, Jorgensen EM (2018) Synaptojanin and endophilin mediate neck formation during ultrafast endocytosis. Neuron 98:1184–1197

Wayne R (1993) Excitability in plant cells. Am Sci 81:140–151

Weinrich M, Worcester DL (2013) Xenon and other volatile anesthetics change domain structure in model lipid raft membranes. J Phys Chem B 117:16141–16147

White RG, Barton DA (2011) The cytoskeleton in plasmodesmata: a role in intercellular transport? J Exp Bot 62:5249–5266

Wildon DC, Thain JF, Minchin PEH, Gubb ER, Reilly AJ, Skipper YD, Doherty HM, O'Donnell PJ, Bowies DJ (1992) Electrical signalling and systemic proteinase inhibitor induction in the wounded plant. Nature 360:62–65

Xue Y, Xing J, Wan Y, Lv X, Fan L, Zhang Y, Song K, Wang L, Wang X, Deng X, Baluška F, Christie JM, Lin J (2018) Arabidopsis blue light receptor phototropin 1 undergoes blue light-induced activation in membrane microdomains. Mol Plant 11:846–859

Yamazaki T, Kawamura Y, Minami A, Uemura M (2008) Calcium-dependent freezing tolerance in Arabidopsis involves membrane resealing via synaptotagmin SYT1. Plant Cell 20:3389–3404

Yamazaki T, Takata N, Uemura M, Kawamura Y (2010) Arabidopsis synaptotagmin SYT1, a type I signal-anchor protein, requires tandem C2 domains for delivery to the plasma membrane. J Biol Chem 285:23165–23176

Yao H, Xu Q, Yuan M (2008) Actin dynamics mediates the changes of calcium level during the pulvinus movement of Mimosa pudica. Plant Signal Behav 3:954–960

Yokawa K, Kagenishi T, Pavlovic A, Gall S, Weiland M, Mancuso S, Baluška F (2018) Anaesthetics stop diverse plant organ movements, affect endocytic vesicle recycling and ROS homeostasis, and block action potentials in Venus flytraps. Ann Bot 122:747–756

Yokawa K, Kagenishi T, Baluška F (2019) Anesthetics, anesthesia, and plants. Trends Plant Sci 24:12–14

Yoshioka T, Takenaka T (1979) Nitellopsis obtusa internodal cell birefringence change during action potential. Biophys Struct Mech 5:1–10

Zhao X, Zhang X, Qu Y, Li R, Baluška F, Wan Y (2015) Mapping of membrane lipid order in root apex zones of Arabidopsis thaliana. Front Plant Sci 6:1151

Zhu J, Bailly A, Zwiewka M, Sovero V, Di Donato M, Ge P, Oehri J, Aryal B, Hao P, Linnert M, Burgardt NI, Lücke C, Weiwad M, Michel M, Weiergräber OH, Pollmann S, Azzarello E, Mancuso S, Ferro N, Fukao Y, Hoffmann C, Wedlich-Söldner R, Friml J, Thomas C, Geisler M (2016) TWISTED DWARF1 mediates the action of auxin transport inhibitors on actin cytoskeleton dynamics. Plant Cell 28:930–948

Ziegler WH, Gingras AR, Critchley DR, Emsley J (2008) Integrin connections to the cytoskeleton through talin and vinculin. Biochem Soc Trans 36:235–239

Chapter 6
The Actomyosin System in Plant Cell Division: Lessons Learned from Microscopy and Pharmacology

Einat Sadot and Elison B. Blancaflor

Abstract Actin and myosin (i.e., the actomyosin system) play pivotal roles in plants, including organelle movement, cytoplasmic streaming, cell expansion, responses to microbes, cell signaling, and cell division. Among the plant biological processes attributed to actin and myosin function, understanding their precise role in cell division has been one of the more challenging problems to address. The difficulties in linking actomyosin function to cell division come in large part from inconsistent actin labeling in the cell division apparatus to the mild cell division phenotypes of actomyosin mutants. While the latter can be explained by functional redundancy, the presence of actin and myosin in the mitotic spindle has been somewhat controversial. Nonetheless, genetically encoded live actin and myosin probes have confirmed some classic microscopy results reported decades ago while also uncovering unique structures associated with the plant cell division machinery. In this chapter, we discuss how early microscopic work and recent live cell imaging data are beginning to provide a more unified view on how the actomyosin system facilitates cell division in plants.

6.1 Introduction

During the mitotic (M) phase of cell division, microtubules (MTs) form special structures that are involved in proper separation of the genetic material between the two daughter cells. While the spindle is a common structure found in animals, fungi, and plants, additional specific microtubule-based structures function in plants. For example, the preprophase band (PPB), a parallel array of cortical MTs that start to appear at the late gap (G) 2 stage, marks the future division site. Another example is the phragmoplast, which forms during telophase. The phragmoplast consists of MTs

E. Sadot (✉)
The Institute of Plant Sciences, Volcani Center, ARO, Rishon-LeZion, Israel
e-mail: vhesadot@volcani.agri.gov.il

E. B. Blancaflor
Noble Research Institute LLC, Ardmore, OK, USA

© Springer Nature Switzerland AG 2019
V. P. Sahi, F. Baluška (eds.), *The Cytoskeleton*, Plant Cell Monographs 24,
https://doi.org/10.1007/978-3-030-33528-1_6

that are perpendicular to the division plane, and it serves as a scaffold for the building of the new cell wall (Wasteneys 2002). The PPB and phragmoplast are well-known MT-based structures that have been described extensively in many basic biology textbooks. Unlike MTs, the role of actin and myosins in the plant cell division apparatus is not as well understood. A major reason for the uncertainty in elucidating the actomyosin system's role in plant cell division might be because filamentous-actin (F-actin) is more difficult to image in plant cells compared to MTs. For example, plant actin was found to be sensitive to aldehyde-based fixatives (Katsuta et al. 1990; Ketelaar et al. 2002; Traas et al. 1987). Therefore, cells should be pretreated with the cross-linker, *m*-maleimidobenzoyl *N*-hydroxysuccinimide ester (MBS), before regular fixation with paraformaldehyde (PFA) or by rapid freezing (Baskin et al. 1996) in order to properly preserve the actin network in many plant cell types. Although green fluorescent protein (GFP)-actin has been expressed in plants, it was unable to consistently form long F-actin (Liu et al. 2004; Lo et al. 2011). Recently, it was found that a long linker of 18 amino acids separating actin and GFP cDNAs is necessary to improve assembly of GFP-actin into long F-actin (Kijima et al. 2018). Therefore, GFP fused to F-actin binding domains of actin regulatory proteins has been the primary tool used to follow actin filaments in living plant cells (Ketelaar et al. 2004; Kost et al. 1998; Riedl et al. 2008; Voigt et al. 2005). However, fluorescent probe-based fusions with actin-binding protein domains might not label the entire population of actin filaments in cells due to competition with endogenous actin regulatory proteins. As such, accurate detection of F-actin in certain cell types might be constrained.

Despite the technical difficulties associated with imaging actin and myosin during plant cell division, there is substantial imaging and pharmacological evidence that they participate in this important process.

Here, we survey the scientific literature addressing the role of the actomyosin system during plant cell division. We focus primarily on early microscopy studies of fixed plant material and actomyosin-disrupting compounds to more recent live cell imaging experiments enabled by fluorescent protein tagging of F-actin and myosin.

6.2 F-actin Is Present in the MT Division Apparatus

6.2.1 Prophase to Anaphase

Co-labeling of MTs and F-actin showed that during mitosis, F-actin remains in the cortex and cytoplasm while MTs are restricted to the division apparatus. However, by using delicate and special labeling methods, evidence started to accumulate during the eighties of the previous century that F-actin can also be observed in the MT division apparatus itself. For example, Traas and colleagues introduced rhodaminyl lysine phallotoxin to carrot cells in suspension by permeabilizing the membrane with detergents or by electroporation. This method allowed the fine visualization of F-actin, and it was shown that a network of F-actin persists in the cytoplasm throughout division, but also associated with the PPB, spindle, and phragmoplast (Traas et al. 1987). During preprophase, when chromatin started to condense, cortical F-actin was observed around the nucleus forming an F-actin PPB-like ring, which in some cells remained there until early metaphase. The presence of F-actin in the spindle was confirmed in Alfalfa (*Medicago sativa* L.) cell suspension cultures stained with rhodamine-labeled phalloidin (Seagull et al. 1987). In these cells, it was possible to detect discrete spindle F-actin during metaphase only when the nucleus was positioned near the surface. In BY2 cells fixed sequentially in MBS and PFA, it was shown that F-actin was associated with the PPB and with MTs in cytoplasmic extensions surrounding the nucleus. After disassembly of the PPB, F-actin remained in cytoplasmic strands. It was proposed that these F-actin networks might be involved in positioning the mitotic apparatus (Katsuta et al. 1990). The F-actin containing cytoplasmic transvacuolar bridge linking the mitotic apparatus to the anticipated division site was observed several decades ago and termed the phragmosome (Sinnott and Bloch 1940).

Information on F-actin in the cell division site was obtained by microinjecting rhodamine phalloidin into living *Tradescantia virginiana* stamen hair cells. This enabled dynamic changes in F-actin organization throughout cell division to be observed (Cleary et al. 1992). It was shown that during chromatin condensation, transverse cortical F-actin bands formed parallel to the MT PPB. The F-actin PPB disappeared before the breakdown of the nuclear envelope, leaving an F-actin-depleted zone at the cortex at the predicted division site. Interestingly, F-actin was found in the cortex and cytoplasm, but remained absent from the cortical division zone (CDZ) throughout mitosis (Cleary et al. 1992).

The presence of F-actin in the PPB was also reported in root tip cells (Palevitz 1987), and cotyledons (Mineyuki and Palevitz 1990) of *Allium cepa*, and in root tip cells of wheat (*Triticum aestivum* L. cv. Kite) (McCurdy and Gunning 1990).

The spatial relationships between F-actin and MTs at different stages of PPB assembly were studied carefully using electron microscopy (Ding et al. 1991) and electron tomography techniques in onion cotyledon epidermal cells (Takeuchi et al. 2016). It was shown that short single F-actin filaments are present among the PPB MTs with some appearing to form bundles of MT-actin-MT bridges. During late prophase, less F-actin filaments were observed at the center of the PPB, which corresponded to the formation of the ADZ, and more pure MT–MT bundles were formed as the PPB narrowed down (Takeuchi et al. 2016).

Taken together, two F-actin structures that formed at late G2 and during mitosis were of particular interest when interpreting image data in dividing cells. First, the F-actin PPB appeared to be wider than the MT PPB (Liu and Palevitz 1992; Mineyuki 1999). Second, the F-actin-depleted zone (ADZ) appeared to result from the disintegration of the central part of the F-actin PPB in late G2. This ADZ persisted throughout mitosis (Cleary et al. 1992) or became less evident during telophase (Liu and Palevitz 1992). The existence of F-actin in the spindle remains controversial since existing live cell F-actin binding GFP probes do not seem to label the spindle although they very prominently mark the phragmoplasts (Wang et al. 2008; Komis et al. 2018; Fig. 6.1).

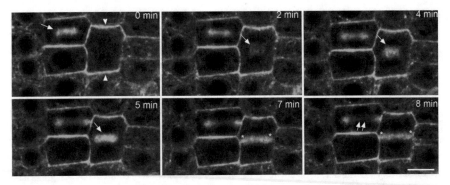

Fig. 6.1 F-actin in the meristematic region of living *Arabidopsis* roots. F-actin is labeled using a GFP-Lifeact construct. Note that F-actin is enriched in phragmoplasts (arrow) and at the end walls of a cell (arrowheads) prior to phragmoplast formation. F-actin in the phragmoplast expands until it reaches the lateral walls (asterisks). As the phragmoplast matures, F-actin signal appears to get weaker along the center (double arrows). Scale bar = 10 μm

Despite inconsistencies in using genetically encoded F-actin binding constructs to mark spindles in living plant cells, they have been useful in confirming other actin-based structures associated with cell division shown in fixed cells. For instance, GFP fused to the second F-actin binding domain of fimbrin (i.e., GFP-ABD2) marked parallel transverse cortical arrays of F-actin during late G2 in Bright-Yellow (BY)-2-GF11 cell lines (Sano et al. 2005). Interestingly, images from the midplane of the cell suggested that while the MT PPB was localized near the cortex, the F-actin PPB was situated on a relatively inner side of the cortex under the MT PPB (Sano et al. 2012). With the beginning of mitosis, this F-actin PPB-like wide band developed a darker stripe in the middle in which less F-actin was observed and two F-actin accumulations flanking it appeared. These structures were named microfilament twin peaks (MFTP) and were observed until metaphase (Sano et al. 2005). At the end of telophase, cell plate fusion occurred at the same ADZ or the center of the MFTP where the presence of F-actin was, at least partially, restored. This same domain marked by the PPB and later by the ADZ is the CDZ. It maintains specific composition of signs for the cell plate and has been described in length in some excellent recent reviews to check if indeed eventually (Muller and Jurgens 2016; Rasmussen and Bellinger 2018; Smertenko et al. 2017). In this respect, it was interesting to note that the BY-2 ADZ was found to be also devoid of kinesin KCA1. The KCA1-depleted zone occurred from metaphase until the end of cytokinesis, but its relation to F-actin remains unclear. The uncertainty as to the relevance of the KCA1-depleted zone for actin-dependent regulation of cell division is because the ADZ disappears if F-actin is disrupted with latrunculin B, but the KCA1-depleted zone is not (Vanstraelen et al. 2006).

An additional role for actin in cell division was described in maize; polarized F-actin patches were found in maize precursor subsidiary mother cells (SMC) that are immediately adjacent to precursor guard mother cell (GMC) (Shao and Dong 2016). It was found that polarization of the WAVE/SCAR complex at the SMC side toward the GMC led to further recruitment of receptor-like kinases, activated ROPs, and the ARP2/3 complex, which led to the formation of the polarized F-actin patch resulting in migration of the nucleus toward that side. This process is required for the subsequent asymmetric cell division leading to the formation of a smaller daughter subsidiary cell adjacent to the stomata (Facette et al. 2015).

In this respect, it is interesting to note that an F-actin probe consisting of ABD2 labeled with GFP at the N- and C-terminus (GFP-ABD2-GFP) or Lifeact-GFP (Fig. 6.1) revealed bright F-actin signals at the apical and basal membranes during telophase prior to phragmoplast formation (Wang et al. 2008). Furthermore, recent light sheet microscopy studies confirmed the existence of these F-actin patches in the apical and basal membranes of root dividing cells during telophase (Komis et al. 2018). The role of end wall-enriched F-actin remains to be determined.

6.2.2 Telophase and Cytokinesis

In late anaphase, fine F-actin was present in the interzone between the two sets of chromosomes. These F-actin networks reorganized to form a phragmoplast-like structure between the reforming nuclei. The F-actin phragmoplast expanded toward the region of the cell cortex that was depleted of F-actin (Cleary et al. 1992). Further, by sequential microinjection of rhodamine- and FITC-labeled phalloidin to *Haemanthus katherinae* endosperm cells during metaphase and anaphase, respectively, it was found that F-actin in the interzone was newly assembled (labeled by FITC) and did not originate from the F-actin encaging the spindle (labeled by rhodamine) (Schmit and Lambert 1990), suggesting separate regulation.

Using BY-2 cells expressing the GFP-fABD2 marker, it was shown that while F-actin was only faintly found in the spindle, fluorescence increased when the phragmoplasts formed (Sano et al. 2005). Microinjection of rhodamine–phalloidin to living stamen hair cells of *Tradescantia* revealed actin filaments that were parallel to each other and to those of MTs (Staehelin and Hepler 1996). Unlike MTs, no overlap between phragmoplast F-actin in the mid zone was detected (Staehelin and Hepler 1996). In late telophase stage of unfixed carrot cells stained with rhodaminyl lysine phallotoxin, it was shown that F-actin is not only associated with the phragmoplast but also bridged the leading edge of the phragmoplast to the cortex (Lloyd and Traas 1988). Based on these observations, it was postulated that the phragmosomal F-actin retains a memory of the division site throughout division (Lloyd and Traas 1988). F-actin bridges linking the phragmoplast and the CDZ were also reported in BY-2 cells labeled with rhodamine-phalloidin (Hasezawa et al. 1994), and in BY-2 cells expressing GFP-fABD2 (Zhang et al. 2009). By expressing EB1-GFP and RFP-ABD2 in BY-2 cells, it was possible to follow the guidance of the phragmoplast by F-actin in live cells. It was shown that F-actin elongating from the growing edge of the phragmoplast to the cell surface eventually coalesces with those found at the cortex (Sano et al. 2012). An elegant experiment was recently reported using BY-2 expressing GFP-fABD2. To better resolve the F-actin bridge between the expanding cell plate and the CDZ, cells were centrifuged, which led to the translocation of the division apparatus away from its position in the center to the edge of the cell. In many imaged cells, F-actin bridges to the CDZ were able to pull the mitotic apparatus at least partially, back to its original place (Arima et al. 2018). It is not clear whether the origin of actin in these bridges is phragmosomal or the phragmoplast or both.

6.3   The Role of F-actin in Cell Division: Insights from Drug Treatments, Actin Binding Proteins, and Mutants

The understanding of the actual role of F-actin during cell division is largely based on experiments using F-actin disrupting drugs. We begin this section by highlighting some of these studies.

In pre-mitotic BY-2 cells, the nucleus migrates from the periphery to the central region of the cell. Treatment with the MT disrupting drug propyzamide or the F-actin disrupting drug cytochalasin D led to only a small decrease in the number of cells with a central nucleus. However, simultaneous treatment with both compounds inhibited nuclear migration to the center of the cell in more than 80% of cells (Katsuta et al. 1990) suggesting a role for both cytoskeletal elements. In subsidiary mother cells (SMCs) of *Tradescantia virginiana*, pre-mitotic nuclei migrate toward the guard mother cell. However, nuclei were unable to migrate in the presence of the F-actin disrupting drug cytochalasin B (Kennard and Cleary 1997). Such observations indicate that F-actin participated in nuclear migration toward the cell division site prior to the initiation of cell division in these cells.

It was shown that treatment of *Allium cepa* seedlings with cytochalasin D resulted in dramatically increased width of MT PPB in both asymmetrical and symmetrical prophase cells (Mineyuki and Palevitz 1990). Inhibition of MT PPB narrowing because of F-actin disruption was confirmed in BY-2 cells expressing GFP-tubulin and treated with 2 μM of latrunculin B (Kojo et al. 2013). Such findings suggest that F-actin plays a role in regulating the narrowing of the MT PPB.

Higher rate of BY-2 cells exhibited abnormal division planes when an F-actin polymerization inhibitor, bistheonellide A (BA), was applied to synchronized cells before the formation of the ADZ but not if it was applied later during cell division. From this work, it was concluded that the loss in "memory" of the future division sites where the ADZ had been located was a major reason for abnormal cell divisions when actin was disrupted (Hoshino et al. 2003). In carrot suspension cells, it was shown that depolymerization of F-actin with cytochalasin D led the spindle axis to reorient 90°, such that it aligned parallel with the phragmosome (Lloyd and Traas 1988). This observation was taken as evidence to support a role for F-actin in the spindle positioning. Guard mother cells (GMC) of *Allium cepa* typically divide longitudinally. However, at the beginning of mitosis, the spindle is oriented obliquely in the cell until telophase, when the growing cell plate undergoes a specific, directional rotation to the longitudinal plane. It was shown that as a result of treatment with the F-actin disrupting drug cytochalasin B, the new cell plate did not rotate from its initial oblique orientation resulting in the formation of abnormally shaped guard cells (Palevitz 1980). Using GFP-fABD2 expressing BY-2 cells, it was shown that while inhibition of F-actin polymerization by BA caused about a 10% decrease in the cell plate expansion rate at the early phase, 25% decrease in cell plate expansion rate occurred at the late phase (Higaki et al. 2008). This suggests that F-actin plays a more significant function at the later stage of cell plate elongation. The decrease in cell plate expansion at the latter phases is in agreement with other findings showing that cell plate expansion can be divided into two phases. The initial stage involves the early expansion of the cell plate until it reaches the width of the daughter nuclei, and the second stage involves the growth of the cell plate until it fuses with the mother cell wall (Valster and Hepler 1997). In addition, it was shown that microinjection of profilin, a G-actin binding protein, to *Tradescantia virginiana* stamen hair cells caused a delay or incomplete cell plate formation (Valster et al. 1997), which again supports a role for F-actin in this process. Interestingly, BA did not block phragmoplast growth, but decreased the rate of its vibrations in relation to the CDZ (i.e., from 15° in control to 5° after BA). At the same time, BA treatment increased the time until the phragmoplast reached 0° from 20 min to 40 min (Sano et al. 2012). The authors assumed that these vibrations helped the phragmoplast F-actin to find their counterpart in the CDZ, a process necessary for the completion of cytokinesis. Contradictory data were found in BY-2 cells in that 1 μM of latrunculin A arrested cytokinesis (Nebenfuhr et al. 2000), but 20 μM of latrunculin B added during metaphase did not interfere with the progression of cytokinesis (Van Damme et al. 2011). Another publication showed that 1 μM latrunculin B applied to cells at metaphase prevented them from entering anaphase and 0.2 μM allowed the progression but led to distorted cell plates (Zhang et al. 2009). These contradictory results showing that disruption of actin at different stages can lead to differential outcomes could potentially be explained by a recent elegant work (van Oostende-Triplet et al. 2017). BY-2 cell plate diameter was plotted over time and was found to form and expand not in two phases (Valster and Hepler 1997) but in three phases. In the first stage, a uniform small disk-shaped region was formed. The second phase is characterized by a rapid expansion, in which the cell plate does not reach the mother cell wall. In the third phase, the cell plate exhibits a slower expansion rate that culminates in its merging with the mother wall at the CDZ (van Oostende-Triplet et al. 2017). Latrunculin B treatment did not arrest the rapid cell plate expansion phase but inhibited the expansion rate during the third phase. It also prevented the cell plate from contacting the mother cell wall, suggesting again a specific role for F-actin at this final stage of cytokinesis. Increased oblique cell plates are also apparent in roots grown continuously in the presence of low concentrations of latrunculin B (Fig. 6.2).

Research on F-actin binding proteins has also expanded our understanding on how F-actin manifests its role in cell division. For example, WLIM1, an F-actin stabilizing protein, was found to co-localize with F-actin in a ring-like structure

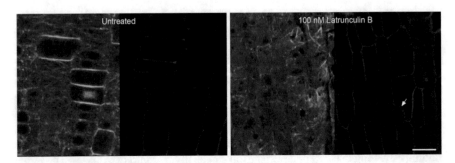

Fig. 6.2 Treatment with low concentrations of latrunculin B disrupts F-actin organization and increases the frequency of obliquely positioned cell plates (arrow) in Arabidopsis roots. Scale bar = 10 μm

during telophase probably stabilizing F-actin in the phragmoplast (Thomas et al. 2006). In other studies, *Arabidopsis* formin FH5 (Ingouff et al. 2005) and *Physcomitrella patens* class II formin (van Gisbergen et al. 2012) were found in the phragmoplast mid zone, suggesting that the barbed ends of phragmoplast F-actin filaments are oriented inward. These observations are in agreement with previous data obtained by electron microscopy in which F-actin filaments decorated with heavy meromyosin (HMM) enabled the visualization of arrowhead-like structures directed toward F-actin pointed ends (Huxley 1963). By isolating phragmoplasts from synchronized BY-2 cells and the identification of their actin filaments by heavy meromyosin arrowhead decoration, it was shown that in about 80% of the filaments, heavy meromyosin arrowheads pointed away from the plate (Kakimoto and Shibaoka 1988). Furthermore, moss formin 2A (For2A-GFP) was shown to be close to the mid zone at the leading edge of the phragmoplast as it expanded, which indicated a role for this protein in local active actin polymerization from the barbed ends (Wu and Bezanilla 2014).

Despite its apparent presence and role in cell division, mutants in actin or in actin-related genes do not exhibit pronounced cell division phenotypes.

The maize *BRK1* and *BRK3* genes encode for subunits of the WAVE/SCARE complex, which are activators of the actin-nucleating ARP2/3 proteins. Their mutants show disruption of division asymmetry in subsidiary guard cells. It was shown that 25% and 10% subsidiary cells are abnormal in *brk1* and *brk3* plants, respectively (Facette et al. 2015).

The *der1–3* mutants, which are disrupted in the vegetative *ACTIN2* gene, displayed higher frequencies of oblique cell division planes and shifted positioning of PPB, and phragmoplasts in root cells (Vaskebova et al. 2018). The mild cell division phenotypes in mutants to actin and actin-regulatory proteins indicate a high degree of functional redundancy in actin-mediated pathways that govern plant cytokinesis.

6.4 The Role of Myosins in Plant Cell Division

Like research on actin's role in cell division, early work to implicate myosin in this process involved the use of chemical inhibitors. For instance, stamen hair cells of *Tradescantia virginiana* L. were treated with 30 mM 2,3-butanedione monoxime (BDM), an inhibitor of myosin ATPase, or 200 mM of ML-7, a specific inhibitor of myosin light chain kinase. In the presence of BDM, most cell plates became tilted, wavy, or fragmented (Molchan et al. 2002). Time-lapse observation showed that while the initiation and early cell plate expansion were normal, the late lateral expansion was inhibited. Phragmoplast actin and MTs lost their perpendicular alignment especially those located at the leading edge (Molchan et al. 2002). Interestingly, ML-7 also mainly inhibited the late lateral expansion of the cell plate (Molchan et al. 2002). Based on studies with inhibitors, it was concluded that myosins are important in propelling and directing the cell plate at the late stage until the completion of cytokinesis but not in motorizing vesicles along the phragmoplast toward the early formed cell plate (Hepler et al. 2002; Molchan et al. 2002).

6.4.1 Myosin VIII

Specific antibodies against the tail domain of myosin VIII ATM1 revealed its subcellular localization in the apical and basal membranes as well as in the newly formed cell plates in the roots of maize and cress (*Lepidium sativum*) (Reichelt et al. 1999). Subsequent GFP fusions confirmed these observations. A screen for cytokinetic proteins by fusion to GFP and expression in BY-2 cells revealed that *Arabidopsis* myosin VIII ATM1 accumulated in the cell plate during telophase (Van Damme et al. 2004). This was further confirmed by expression of the full-length cDNA of ATM1 fused to GFP at its N-terminus, under the regulation of its native promoter, in the Arabidopsis *atm1* mutant background. It was shown in these plants that GFP-ATM1 localized to newly formed cell plates in the roots (Haraguchi et al. 2014). In the moss, it was shown that in Δmyo8ABCDE plants, a high percentage of cell plates were aberrantly positioned with respect to the filament axis in both branching and non-branching cells (Wu and Bezanilla 2014). Expressing Myo8A-GFP in the quintuple mutant partially restored cell plate positioning. In branching cells, GFP fluorescence of the Myo8A-GFP construct appeared as a ring in the cortex prior to prophase. This was followed by Myo8A-GFP accumulation at the spindle and phragmoplast. During phragmoplast expansion, Myo8A-GFP was found in its periphery. The accumulation of Myo8A-GFP created the formation of two rings: an inner ring on the phragmoplast and an outer ring at the cell cortex. Of note, moss does not have MT PPBs (Wu and Bezanilla 2014). To test the hypothesis that moss Myo8A might function as a PPB, it was expressed in BY-2 cells. It was found that in prophase, cortical Myo8A-GFP formed a PPB-like band, corresponding to the CDZ, which persisted until the end of cytokinesis. In early cytokinesis, Myo8A-GFP also appeared at the phragmoplast mid zone. In BY2 cells, moss Myo8A-GFP followed the expansion of the phragmoplast, eventually reaching the cortical division site resulting in the formation of two rings. Further, it was shown that by binding to their plus ends, Myo8 directed peripheral phragmoplast MTs to the cortical division site in an actin-dependent manner (Wu and Bezanilla 2014).

6.4.2   Myosin XI

The presence of myosin XI in the cell division apparatus was first reported in 2009 when BY-2 cells were stained with specific antibodies prepared against the 175 kDa myosin protein isolated and characterized from tobacco (Yokota et al. 2009). It was shown that at preprophase, the signal corresponding to 175 kDa myosin was found in the PPB. In metaphase cells, myosin XI was localized in the cytoplasm and the spindle region. During telophase, myosin accumulated in the mid zone of the phragmoplast and around daughter nuclei. Cytoplasmic strands also contained the 175 kDa myosin signal. Double staining of actin and the 175 kDa myosin showed an overlap between the two. Interestingly, ER staining also overlapped with that of the 175 kDa myosin and the two co-fractionated on a sucrose gradient. While BDM did not inhibit the presence of ER in the spindle, it did suppress the presence of the ER in the phragmoplast mid zone and around the nuclei during telophase, again indicating a specific role of the actomyosin system at this stage. Nuclear migration to the center of daughter cells was inhibited in the presence of BDM. This observation is consistent with data showing myosin XI-I localization around the nucleus (Avisar et al. 2009; Tamura et al. 2013), and its role in regulating nuclear migration (Tamura et al. 2013).

In a recent study, *Physcomitrella patens* myosin XIa was fused to 3mGFP at its C-terminus (Sun et al. 2018). Myosin XIa-GFP was found associated with the spindle and the cell plate. Latrunculin B caused a reduced myosin XIa-GFP signal in the cell plate, but such a reduction was minimal in the spindle. To understand the relationships between the accumulation of myosin XIa and vesicles in the cell plates, fluorescence of 3mGFP-VAMP 722 was compared to that of myosin XIa-3mGFP. It was found that while both signals accumulated at the cell plate, VAMP-722 was independent of actin. On the basis of this observation, it was assumed that the delivery of VAMP-722 vesicles to the cell plate was mainly dependent on microtubules (Sun et al. 2018). A comparison of myosin VIII and myosin XIa revealed no major differences between the accumulation of both in the spindle and cell plate of *P. patens* caulonemal cells.

When *Arabidopsis* myosin XIK-YFP was expressed under its native promoter in the triple mutant *xi1/xi2/xik*, it was shown to be functional (Peremyslov et al. 2010). Interestingly, XIK-YFP was found in the spindle and phragmoplast mid zone in dividing cells in both root and shoot apical meristems (Abu-Abied et al. 2018). In the cell plate, XIK-YFP formed a disk at early telophase and then accumulated at the cell plate leading edge, forming a ring until cytokinesis had completed. Expressing the MT marker mCherry MAP 4 MBD in the *xi1/xi2/xik* triple mutant (3KO) and in the recovered line (3KO + YFP-XI-K) enables the timing of mitosis to be quantified. It was shown that cell division was delayed in the 3KO plants (65 $\pm$ 8.5 min) compared to wild-type (39 $\pm$ 7 min) or the 3KOR plants (45 $\pm$ 11 min). Interestingly, YFP-XIK appeared transiently at the CDZ in metaphase, disappeared during anaphase and early telophase, and returned during late telophase (Abu-Abied et al. 2018). The pattern of YFP-XIK signals during these various stages suggests a role for myosin XIK in cell division orientation control. Higher frequency of oblique cell division planes was detected in the stele of the 3KO plants (Abu-Abied et al. 2018). Importantly, two major myosin XIK interactors MyoB1 and MyoB2 were undetectable in dividing cells, suggesting that other myosin binding proteins might function in coordination with myosin XIK during cell division.

6.5 Conclusions and Perspectives

Bulk of our knowledge about the role of the actomyosin system in plant cell division has come primarily from microscopy and pharmacological approaches. Such approaches indicate that actin and myosin are present in the plant cell division apparatus from early prophase to late telophase (Fig. 6.3), and disrupting their function with chemical inhibitors delays the progress of cell division or the proper positioning of the cell plate. Like in yeast and animals, the molecular machinery that drives cell division in plants is robust and involves redundant pathways. With regard to the actomyosin system, plant genomes encode for several actin and myosin proteins. Therefore, when one or more actin or myosin genes are deleted, remaining genes still allow the plant to complete cell division with only mild growth phenotypes (Vaskebova et al. 2018; Abu-Abied et al. 2018).

The redundant and complex pathways that underlie plant cell division have made it challenging to uncover the precise mechanisms by which the actomyosin system participates in plant cell division. Dividing cells of *Physcomitrella patens* which lack PPB but still contain several conserved components for marking the site of cell division might serve as a good system to investigate the contribution of the actomyosin system (Wu et al. 2018). In addition to new genetic systems, microscopy and pharmacology have to be complemented with proteomic and biochemical approaches. Such approaches would help in expanding the inventory of plant proteins that comprise structures associated with cell division. For instance, proteomics of isolated mammalian midbodies led to the identification of several cytoskeletal and membrane proteins crucial for cytokinesis (Skop et al. 2004). Revisiting and refining methods to isolate PPBs and/or phragmoplasts (Kakimoto and Shibaoka 1988) or known proteins (Rybak et al. 2014) for downstream proteomic and biochemical analyses would be an important advance for the field.

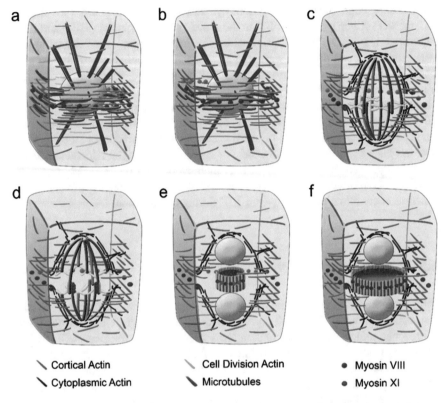

Cortical Actin

Cytoplasmic Actin

Cell Division Actin

Microtubules

● Myosin VIII

● Myosin XI

Fig. 6.3 F-actin and myosins are associated with various structures relevant to plant cell division. (**a**) Early prophase, actin forms a wider PPB, than that of MTs (Cleary et al. 1992; Ding et al. 1991; Katsuta et al. 1990; Liu and Palevitz 1992; McCurdy and Gunning 1990; Mineyuki 1999; Mineyuki and Palevitz 1990; Palevitz 1987; Sano et al. 2005, 2012; Takeuchi et al. 2016), regulating the narrowing of the MT PPB and is found in cytoplasmic strands positioning the nucleus in the anticipated cell division zone (Katsuta et al. 1990). Moss myosin VIII forms a ring in the absence of MT PPB (Wu and Bezanilla 2014). (**b**) Late prophase, MT PPB narrows down and actin is depleted from the central cortex creating the actin-depleted zone ADZ (Cleary et al. 1992; Hoshino et al. 2003; Liu and Palevitz 1992; Sano et al. 2005; Takeuchi et al. 2016) contributing to the memory of the cell division site. (**c**) Metaphase, PPB disassembles, MTs disappear from the cytoplasmic strands, the spindle forms, actin remains in the cortex and cytoplasm strands (Katsuta et al. 1990; Sano et al. 2005), and forms a basket encaging the spindle (Schmit and Lambert 1990) contributing to its positioning. Myosin XIK transiently appears in the CDZ (Abu-Abied et al. 2018). (**d**) Anaphase, actin (Cleary et al. 1992; Schmit and Lambert 1990), myosin VIII (Wu and Bezanilla 2014), and myosin XIK (Abu-Abied et al. 2018) appear in the spindle mid zone. (**e**) Early telophase, actin appears in the phragmoplast (Cleary et al. 1992; Sano et al. 2005; Schmit and Lambert 1990; Staehelin and Hepler 1996; Zhang et al. 2009), and also encages the phragmoplast including the nuclei (Hasezawa et al. 1994; Lloyd and Traas 1988), myosin XIK is present throughout the forming cell plate (Abu-Abied et al. 2018), and myosin VIII forms two rings, in the cortex and the phragmoplast. (**f**) Late telophase, actin is present in the extending phragmoplast, and also forms a bridge to the CDZ (Arima et al. 2018; Hasezawa et al. 1994; Lloyd and Traas 1988; Sano et al. 2012; Zhang et al. 2009). Myosin VIII at the expanding phragmoplast and cortex. Myosin XI concentrates at the cell plate edge and reappears in the CDZ (Abu-Abied et al. 2018). During telophase, the actomyosin system contributes to positioning and orienting the phragmoplast to the CDZ, and promoting the cell plate expansion at the last phase until the fusion with the mother cell wall

Acknowledgment Research on the plant cytoskeleton in the authors' laboratories is supported by the National Aeronautics and Space Administration (NASA grant numbers 80NSSC18K1462 and 80NSSC19KO129) to EBB and BSF Binational Science Foundation (grant number 2013084) to ES.

References

Abu-Abied M, Belausov E, Hagay S, Peremyslov V, Dolja V, Sadot E (2018) Myosin XI-K is involved in root organogenesis, polar auxin transport, and cell division. J Exp Bot 69:2869–2881

Arima K, Tamaoki D, Mineyuki Y, Yasuhara H, Nakai T, Shimmen T, Yoshihisa T, Sonobe S (2018) Displacement of the mitotic apparatuses by centrifugation reveals cortical actin organization during cytokinesis in cultured tobacco BY-2 cells. J Plant Res 131(5):803–815

Avisar D, Abu-Abied M, Belausov E, Sadot E, Hawes C, Sparkes IA (2009) A comparative study of the involvement of 17 Arabidopsis myosin family members on the motility of Golgi and other organelles. Plant Physiol 150:700–709

Baskin TI, Miller DD, Vos JW, Wilson JE, Hepler PK (1996) Cryofixing single cells and multicellular specimens enhances structure and immunocytochemistry for light microscopy. J Microsc 182:149–161

Cleary AL, Gunning BES, Wasteneys GO, Hepler PK (1992) Microtubule and F-actin dynamics at the division site in living Tradescantia stamen hair cells. J Cell Sci 103:977–988

Ding B, Turgeon R, Parthasarathy MV (1991) Microfilaments in the preprophase band of freeze substituted tobacco root cells. Protoplasma 165:209–211

Facette MR, Park Y, Sutimantanapi D, Luo A, Cartwright HN, Yang B, Bennett EJ, Sylvester AW, Smith LG (2015) The SCAR/WAVE complex polarizes PAN receptors and promotes division asymmetry in maize. Nat Plants 1:14024

Haraguchi T, Tominaga M, Matsumoto R, Sato K, Nakano A, Yamamoto K, Ito K (2014) Molecular characterization and subcellular localization of Arabidopsis class VIII myosin, ATM1. J Biol Chem 289:12343–12355

Hasezawa S, Sano T, Nagata T (1994) Oblique cell plate formation in tobacco BY-2 cells originates in double preprophase bands. J Plant Res 107:355–359

Hepler PK, Valster A, Molchan T, Vos JW (2002) Roles for kinesin and myosin during cytokinesis. Philos Trans R Soc Lond Ser B Biol Sci 357:761–766

Higaki T, Kutsuna N, Sano T, Hasezawa S (2008) Quantitative analysis of changes in actin microfilament contribution to cell plate development in plant cytokinesis. BMC Plant Biol 8:80

Hoshino H, Yoneda A, Kumagai F, Hasezawa S (2003) Roles of actin-depleted zone and preprophase band in determining the division site of higher-plant cells, a tobacco BY-2 cell line expressing GFP-tubulin. Protoplasma 222:157–165

Huxley HE (1963) Electron microscope studies on the structure of natural and synthetic protein filaments from striated muscle. J Mol Biol 7:281–IN230

Ingouff M, Fitz Gerald JN, Guerin C, Robert H, Sorensen MB, Van Damme D, Geelen D, Blanchoin L, Berger F (2005) Plant formin AtFH5 is an evolutionarily conserved actin nucleator involved in cytokinesis. Nat Cell Biol 7:374–380

Kakimoto T, Shibaoka H (1988) Cytoskeletal ultrastructure of phragmoplast-nuclei complexes isolated from cultured tobacco cells. In: Tazawa M (ed) Cell dynamics protoplasma (Supplementum 2), vol 2. Springer, Vienna, pp 95–103

Katsuta J, Hashiguchi Y, Shibaoka H (1990) The role of the cytoskeleton in positioning of the nucleus in premitotic tobacco BY-2 cells. J Cell Sci 95:413–422

Kennard JL, Cleary AL (1997) Pre-mitotic nuclear migration in subsidiary mother cells of Tradescantia occurs in G1 of the cell cycle and requires F-actin. Cell Motil Cytoskeleton 36:55–67

Ketelaar T, Faivre-Moskalenko C, Esseling JJ, de Ruijter NC, Grierson CS, Dogterom M, Emons AM (2002) Positioning of nuclei in Arabidopsis root hairs: an actin-regulated process of tip growth. Plant Cell 14:2941–2955

Ketelaar T, Allwood EG, Anthony R, Voigt B, Menzel D, Hussey PJ (2004) The actin-interacting protein AIP1 is essential for actin organization and plant development. Curr Biol 14:145–149

Kijima ST, Staiger CJ, Katoh K, Nagasaki A, Ito K, Uyeda TQP (2018) Arabidopsis vegetative actin isoforms, AtACT2 and AtACT7, generate distinct filament arrays in living plant cells. Sci Rep 8:4381–4381

Kojo KH, Higaki T, Kutsuna N, Yoshida Y, Yasuhara H, Hasezawa S (2013) Roles of cortical actin microfilament patterning in division plane orientation in plants. Plant Cell Physiol 54:1491–1503

Komis G, Novak D, Ovecka M, Samajova O, Samaj J (2018) Advances in imaging plant cell dynamics. Plant Physiol 176:80–93

Kost B, Spielhofer P, Chua NH (1998) A GFP-mouse talin fusion protein labels plant actin filaments in vivo and visualizes the actin cytoskeleton in growing pollen tubes. Plant J 16:393–401

Liu B, Palevitz BA (1992) Organization of cortical microfilaments in dividing root cells. Cell Motil Cytoskeleton 23:252–264

Liu AX, Zhang SB, Xu XJ, Ren DT, Liu GQ (2004) Soluble expression and characterization of a GFP-fused pea actin isoform (PEAc1). Cell Res 14:407–414

Lloyd CW, Traas JA (1988) The role of F-actin in determining the division plane of carrot suspension cells. Drug studies. Development 102:211–221

Lo YS, Cheng N, Hsiao LJ, Annamalai A, Jauh GY, Wen TN, Dai H, Chiang KS (2011) Actin in mung bean mitochondria and implications for its function. Plant Cell 23:3727–3744

McCurdy DW, Gunning BES (1990) Reorganization of cortical actin microfilaments and microtubules at preprophase and mitosis in wheat root-tip cells: a double label immunofluorescence study. Cell Motil Cytoskeleton 15:76–87

Mineyuki Y (1999) The preprophase band of microtubules: its function as a cytokinetic apparatus in higher plants. Int Rev Cytol 187:1–49

Mineyuki Y, Palevitz BA (1990) Relationship between preprophase band organization, F-actin and the division site in Allium; fluorescence and morphometric studies on cytochalasin-treated cells. J Cell Sci 97:283–295

Molchan TM, Valster AH, Hepler PK (2002) Actomyosin promotes cell plate alignment and late lateral expansion in Tradescantia stamen hair cells. Planta 214:683–693

Muller S, Jurgens G (2016) Plant cytokinesis-no ring, no constriction but centrifugal construction of the partitioning membrane. Semin Cell Dev Biol 53:10–18

Nebenfuhr A, Frohlick JA, Staehelin LA (2000) Redistribution of Golgi stacks and other organelles during mitosis and cytokinesis in plant cells. Plant Physiol 124:135–151

Palevitz BA (1980) Comparative effects of phalloidin and cytochalasin B on motility and morphogenesis in Allium. Can J Bot 58:773–785

Palevitz BA (1987) Actin in the preprophase band of *Allium cepa*. J Cell Biol 104:1515–1519

Peremyslov VV, Prokhnevsky AI, Dolja VV (2010) Class XI myosins are required for development, cell expansion, and F-actin organization in *Arabidopsis*. Plant Cell 22:1881–1897

Rasmussen CG, Bellinger M (2018) An overview of plant division-plane orientation. New Phytol 219:505–512

Reichelt S, Knight AE, Hodge TP, Baluška F, Samaj J, Volkmann D, Kendrick-Jones J (1999) Characterization of the unconventional myosin VIII in plant cells and its localization at the post-cytokinetic cell wall. Plant J 19:555–567

Riedl J, Crevenna AH, Kessenbrock K, Yu JH, Neukirchen D, Bista M, Bradke F, Jenne D, Holak TA, Werb Z, Sixt M, Wedlich-Soldner R (2008) Lifeact: a versatile marker to visualize F-actin. Nat Methods 5:605–607

Rybak K, Steiner A, Synek L, Klaeger S, Kulich I, Facher E, Wanner G, Kuster B, Zarsky V, Persson S, Assaad FF (2014) Plant cytokinesis is orchestrated by the sequential action of the TRAPPII and exocyst tethering complexes. Dev Cell 29:607–620

Sano T, Higaki T, Oda Y, Hayashi T, Hasezawa S (2005) Appearance of actin microfilament 'twin peaks' in mitosis and their function in cell plate formation, as visualized in tobacco BY-2 cells expressing GFP-fimbrin. Plant J 44:595–605

Sano T, Hayashi T, Kutsuna N, Nagata T, Hasezawa S (2012) Role of actin microfilaments in phragmoplast guidance to the cortical division zone. Curr Top Plant Biol 13:87–94

Schmit AC, Lambert AM (1990) Microinjected fluorescent phalloidin in vivo reveals the F-actin dynamics and assembly in higher plant mitotic cells. Plant Cell 2:129–138

Seagull RW, Falconer MM, Weerdenburg CA (1987) Microfilaments: dynamic arrays in higher plant cells. J Cell Biol 104:995–1004

Shao W, Dong J (2016) Polarity in plant asymmetric cell division: division orientation and cell fate differentiation. Dev Biol 419:121–131

Sinnott EW, Bloch R (1940) Cytoplasmic behavior during division of vacuolate plant cells. Proc Natl Acad Sci USA 26:223–227

Skop AR, Liu H, Yates J 3rd, Meyer BJ, Heald R (2004) Dissection of the mammalian midbody proteome reveals conserved cytokinesis mechanisms. Science 305:61–66

Smertenko A, Assaad F, Baluska F, Bezanilla M, Buschmann H, Drakakaki G, Hauser MT, Janson M, Mineyuki Y, Moore I, Muller S, Murata T, Otegui MS, Panteris E, Rasmussen C, Schmit AC, Samaj J, Samuels L, Staehelin LA, Van Damme D, Wasteneys G, Zarsky V (2017) Plant cytokinesis: terminology for structures and processes. Trends Cell Biol 27:885–894

Staehelin LA, Hepler PK (1996) Cytokinesis in higher plants. Cell 84:821–824

Sun H, Furt F, Vidali L (2018) Myosin XI localizes at the mitotic spindle and along the cell plate during plant cell division in *Physcomitrella patens*. Biochem Biophys Res Commun 506 (2):409–421

Takeuchi M, Karahara I, Kajimura N, Takaoka A, Murata K, Misaki K, Yonemura S, Staehelin LA, Mineyuki Y (2016) Single microfilaments mediate the early steps of microtubule bundling during preprophase band formation in onion cotyledon epidermal cells. Mol Biol Cell 27:1809–1820

Tamura K, Iwabuchi K, Fukao Y, Kondo M, Okamoto K, Ueda H, Nishimura M, Hara-Nishimura I (2013) Myosin XI-i links the nuclear membrane to the cytoskeleton to control nuclear movement and shape in *Arabidopsis*. Curr Biol 23:1776–1781

Thomas C, Hoffmann C, Dieterle M, Van Troys M, Ampe C, Steinmetz A (2006) Tobacco WLIM1 is a novel F-actin binding protein involved in actin cytoskeleton remodeling. Plant Cell 18:2194–2206

Traas JA, Doonan JH, Rawlins DJ, Shaw PJ, Watts J, Lloyd CW (1987) An actin network is present in the cytoplasm throughout the cell cycle of carrot cells and associates with the dividing nucleus. J Cell Biol 105:387–395

Valster AH, Hepler PK (1997) Caffeine inhibition of cytokinesis: effect on the phragmoplast cytoskeleton in living Tradescantia stamen hair cells. Protoplasma 196:155–166

Valster AH, Pierson ES, Valenta R, Hepler PK, Emons A (1997) Probing the plant actin cytoskeleton during cytokinesis and interphase by profilin microinjection. Plant Cell 9:1815–1824

Van Damme D, Bouget FY, Van Poucke K, Inze D, Geelen D (2004) Molecular dissection of plant cytokinesis and phragmoplast structure: a survey of GFP-tagged proteins. Plant J 40:386–398

Van Damme D, Gadeyne A, Vanstraelen M, Inze D, Van Montagu MC, De Jaeger G, Russinova E, Geelen D (2011) Adaptin-like protein TPLATE and clathrin recruitment during plant somatic cytokinesis occurs via two distinct pathways. Proc Natl Acad Sci USA 108:615–620

van Gisbergen PA, Li M, Wu SZ, Bezanilla M (2012) Class II formin targeting to the cell cortex by binding PI(3,5)P(2) is essential for polarized growth. J Cell Biol 198:235–250

van Oostende-Triplet C, Guillet D, Triplet T, Pandzic E, Wiseman PW, Geitmann A (2017) Vesicle dynamics during plant cell cytokinesis reveals distinct developmental phases. Plant Physiol 174:1544–1558

Vanstraelen M, Van Damme D, De Rycke R, Mylle E, Inze D, Geelen D (2006) Cell cycle-dependent targeting of a kinesin at the plasma membrane demarcates the division site in plant cells. Curr Biol 16:308–314

Vaskebova L, Samaj J, Ovecka M (2018) Single-point ACT2 gene mutation in the Arabidopsis root hair mutant der1-3 affects overall actin organization, root growth and plant development. Ann Bot 122:889–901

Voigt B, Timmers AC, Samaj J, Muller J, Baluška F, Menzel D (2005) GFP-FABD2 fusion construct allows in vivo visualization of the dynamic actin cytoskeleton in all cells of Arabidopsis seedlings. Eur J Cell Biol 84:595–608

Wang YS, Yoo CM, Blancaflor EB (2008) Improved imaging of actin filaments in transgenic Arabidopsis plants expressing a green fluorescent protein fusion to the C- and N-termini of the fimbrin actin-binding domain 2. New Phytol 177:525–536

Wasteneys GO (2002) Microtubule organization in the green kingdom: chaos or self-order? J Cell Sci 115:1345–1354

Wu SZ, Bezanilla M (2014) Myosin VIII associates with microtubule ends and together with actin plays a role in guiding plant cell division. elife 3:e03498

Wu SZ, Yamada M, Mallett DR, Bezanilla M (2018) Cytoskeletal discoveries in the plant lineage using the moss *Physcomitrella patens*. Biophys Rev 10:1683–1693

Yokota E, Ueda S, Tamura K, Orii H, Uchi S, Sonobe S, Hara-Nishimura I, Shimmen T (2009) An isoform of myosin XI is responsible for the translocation of endoplasmic reticulum in tobacco cultured BY-2 cells. J Exp Bot 60:197–212

Zhang Y, Zhang W, Baluska F, Menzel D, Ren H (2009) Dynamics and roles of phragmoplast microfilaments in cell plate formation during cytokinesis of tobacco BY-2 cells. Chin Sci Bull 54:2051–2061

Chapter 7
Cooperation Between Auxin and Actin During the Process of Plant Polar Growth

Jie Liu and Markus Geisler

Abstract Polar growth is provided by rapid cell expansion that spatially focuses at the tip. The regulation and maintenance of polar growth requires two important intracellular events: intensive exocytosis in the tip region and a highly dynamic cytoskeleton system. The selective transport of secretory vesicles and their accumulation in the apical region, which is driven by motor proteins that move along actin cables, is critical for plant polar growth. The regulation of vesicle trafficking and actin cytoskeleton turnover is affected by several intracellular components and signaling pathways. Auxin as one of the most important intracellular elements regulating plant growth and development involved in numerous signaling pathways that can affect the organization and dynamics of cell cytoskeleton is a potential candidate as a polar growth regulator. The current knowledge is summarized here in order to highlight the role of auxin in plant polar growth regulation and the cooperation between auxin and the actin cytoskeleton during this process.

7.1 Introduction

Polar growth, also known as tip growth, is a spatially focused cell expansion that exists in a few types of cells, including hyphae in fungi, pollen tubes and root hairs in plants, and neurites in animals (Pierson et al. 1996; Galway et al. 1997; Gomez and Spitzer 1999; Geitmann and Emons 2000). Root hairs develop at hair-forming cells of the plant root epidermis in order to increase the surface area for absorption of water and nutrients, to support the symbiotic interaction with soil microorganisms. They are single cells and are rarely branched (Gilroy and Jones 2000). Pollen tubes develop from the germinating pollen grains and deliver sperm cells from pollen grains to the ovule through the female tissue of the pistil for double fertilization (Taylor and Hepler 1997; Lord 2000). Pollen tubes and root hairs are different in origin, function, and growth rate but share a common polarity growth pattern (Hepler et al. 2001). In the two cell types, all growth is focused on the specific tip region.

J. Liu · M. Geisler (✉)
Department of Biology, University of Fribourg, Fribourg, Switzerland
e-mail: markus.geisler@unifr.ch

© Springer Nature Switzerland AG 2019
V. P. Sahi, F. Baluška (eds.), *The Cytoskeleton*, Plant Cell Monographs 24,
https://doi.org/10.1007/978-3-030-33528-1_7

Components used for new plasma membrane and cell wall formation are delivered to the growth region by secretory vesicles, trafficking along highly dynamic actin filament framework in the apex. The process of polar growth involves membrane trafficking and polarized actin organization, which are regulated by several singling pathways involving Rho/Rac of plants (ROP) GTPases singling pathway, reactive oxygen species (ROS), phosphoinositides, and Ca^{2+}-dependent protein kinase signaling pathways (Hepler et al. 2001; Feijó et al. 2004; Monteiro et al. 2005; Smith and Oppenheimer 2005). These key components and signaling pathways cross talk form a highly interconnected network that coordinates the cellular activities required for the regulation of polar growth.

Polar growth is an extreme growth format with an astonishing speed compared with the conventional cell elongation: Root hair growth rate is around 10–40 nm/s (Galway et al. 1997), tobacco pollen tube growth rate is around 80–100 nm/s (Derksen et al. 1995), the average *Arabidopsis* pollen growth rate is around 80 nm/s (Boavida and McCormick 2007), while lily pollen growth is faster and can reach 250 nm/s in vitro (Messerli and Robinson 1997). This high-speed polar growth rate correlates with distinct structure and cytoplasmic transport components, as well as highly with a dynamic cytoskeleton network in the tip region of the pollen tube and root hair. In pollen tubes, the polarized cytoplasm is divided into distal region and viable streaming region. The dynamic streaming region is further subdivided into organelles rich shank, subapical zone, and apical zone (the so-called clear zone). The apical zone is a domain absent of organelles, which is occupied exclusively by exocytic vesicles (Yang 1998; Cheung and Wu 2007). However, the "clear zone" typically disappears in nongrowing tubes (Hepler and Winship 2015). In different subregions, F-actin forms different structures that play an essential role in vesicle trafficking. In the shank region, the longitudinal actin cables formation can be induced by the *Arabidopsis thaliana* formin3 (AtFH3) (Ye et al. 2009). The bundled actin cable supporting the cytoplasmic organization by providing actin tracks for vesicle trafficking is essential for the reverse fountain pattern cytoplasmic streaming (Fu et al. 2001; Ye et al. 2009). The subapical zone contains a densely arranged, highly dynamic cortical actin filament structure, the actin fringe, which can capture the vesicles released from the actin cables. Exocytic and endocytic are highly activated in subapical zone, supporting the tip-direction plasma membrane and cell wall components transport (Lovy-Wheeler et al. 2005; Kroeger et al. 2009; Hepler and Winship 2015; Stephan 2017). Several actin-binding proteins (ABPs), including fimbrins, villins, profilins, and formins, have been reported to participate in the maintenance of the actin fringe (Cheung et al. 2010; Wu et al. 2010b; Su et al. 2012b; Qu et al. 2013; Liu et al. 2015; Li et al. 2017). Root hairs exhibit similar polarized characteristics as pollen tube, except a few visible differences: actin filaments in the subapical region are more diffused and extend further into the apical region; a few organelles are observed in the apical region (Hepler et al. 2001; Carol and Dolan 2002). Although the core properties, features, and function of actin structures are not exactly the same in several cell types of diverse species, it is clear that the overall organization and dynamics of actin filaments play an essential role in the maintenance of rapid plant polar growth.

The plant hormone auxin, a key regulator of plant growth and development, is well known to control the directional growth at the organ and tissue level. Auxin is widely distributed in reproductive tissues, including pollen and root hair, shown to be essential for pollen and root hair development and cell elongation (Pitts et al. 1998; Feng et al. 2006; Cecchetti et al. 2008; Jones et al. 2009; Koltai et al. 2010; Sakata et al. 2010; Ding et al. 2012). Low concentration of a synthetic auxin stimulates pollen tube growth in vitro (Chen and Zhao 2008) and root hair elongation (Pitts et al. 1998). Although the signaling mechanisms remain unclear, several shreds of evidence support the idea that auxin may regulate plant polar growth and vesicle trafficking by the PIN-auxin-ROP signaling module (Xu et al. 2010; Bosco et al. 2012; Ding et al. 2012; Nagawa et al. 2012). In addition, auxin can also affect actin expression, induce actin reorganization, and modulate actin dynamics, which feedbacks on polar auxin transport (Kandasamy et al. 2001; Maisch and Nick 2007; Nick et al. 2009; Durst et al. 2013; Li et al. 2014b; Zhu et al. 2016; Ding et al. 2018b). It has thus been proposed that the action of auxin and that of the actin cytoskeleton are tightly interconnected (Zhu and Geisler 2015). All pieces of evidence suggest that auxin is a potential candidate regulating the signal transduction pathway of plant polar growth. Here we review the current literature in order to highlight the role of auxin and actin interaction in the process of plant polar growth.

7.2   The Cytoskeleton in Polar Growth of Cells

The directed or polar distribution of cellular contents provided by vesicle trafficking is the basis of polar growth (Derksen et al. 1995; Hepler and Winship 2015; Pan et al. 2015). The organelle and vesicle movement is driven by motor proteins that move polarly along specific actin filament bundles in a process that requires ATP hydrolysis (Quintana-Cataño et al. 2016; Duan et al. 2018; Nebenführ and Dixit 2018). Therefore, it is necessary to understand the structure and kinetic characteristics of the cytoskeleton in order to understand this membrane transport process in polar growth.

The cytoskeleton of polar growing cells comprises actin filaments and microtubules that mainly function in vesicles trafficking and strengthen the cell wall behind the apical growing region and thus keep the cylindrical shape of the cells. The movement of vesicles and organelles is mainly based on actin filaments, while the role of microtubules is less characterized.

7.2.1   Role of Microtubules in Polar Growth

Microtubules are proposed to be involved in the assembly of cell walls, the movement of sperm cell and short-range vesicles, and organelle trafficking (Raudaskoski et al. 2001; Laitiainen et al. 2002; Romagnoli et al. 2003). In the shank region, long and axially organized thick microtubules are found, sometimes with a helical organization (Traas et al. 1985; Del Casino et al. 1993). At the apical and subapical regions, the organization and function of microtubules are poorly understood due to limitations in chemical fixation techniques. Improved immobilization methods have revealed a tubulin fringe structure in the subapical region, but till now the presence of the tubulin fringe remains controversial. Also, the putative function of this structure and its relationship with the actin fringe is not known (Bibikova et al. 1999; Lovy-Wheeler et al. 2005). Depolymerization of microtubules with oryzalin results in straighter pollen tubes and waving root hair, while stabilization of microtubules with taxol promotes the formation of multiple apical tips in root hairs, indicating that microtubules affect polar growth direction but not its growth rate (Bibikova et al. 1999; Gossot and Geitmann 2007).

Many microtubule-associated proteins (MAPs), reported to regulate cortical microtubule organization and dynamics in vegetative tissues, also exist in pollen tubes. But recent studies suggest that these MAPs (including MAP 18, MDP25, and RIC1) prefer to regulate actin filament dynamic rather than to modulate microtubule turnover during polar growth of pollen tubes. All those three proteins exhibit a Ca^{2+} dependent F-actin severing activity in vitro (Zhu et al. 2013; Qin et al. 2014; Zhou et al. 2015). All these findings suggest that actin plays an essential role during polar growth, while the function of microtubules is subsidiary, and seems to lie in the interaction with the actin cytoskeleton enabling to better negotiate the polar growth paths.

7.2.2 Role of Actin Microfilaments in Polar Growth

Pharmacology combined with live cell imaging has revealed that fine actin filaments are essential for polar growth and for regulating cellular trafficking. In the subapical region of the lily pollen tube, an actin fringe can be easily observed by phalloidin staining (Foissner et al. 2002). However, improved labeling techniques using actin markers, including mTalin, fABD2, and Lifeact, revealed subtler actin structure and dynamics, like the actin nucleation that can be seen initially from the apical and flanking membrane (Cheung and Wu 2004; Sheahan et al. 2004; Voigt et al. 2005; Riedl et al. 2008; Vidali et al. 2009a; Li et al. 2015). Distinct forms of actin filaments in different subcellular regions carry out specific functions in the pollen tube. Disrupting actin organization and dynamics with high concentrations of the actin inhibitor, latrunculin B (Lat B), inhibits cytoplasmic streaming moving along longitudinal actin cables in the shank region. This provided strong evidence for an involvement of the actin cytoskeleton system in long-range, intracellular motility (Andersland et al. 1994; Vidali and Hepler 1997; Hepler et al. 2001). With low concentrations of Lat B, which did not affect the longitudinal actin bundles in the shank region, secretory vesicles rapidly dissipated in tip inverted cone. After inhibitor washout, actin fringe reassembled, secretory vesicles reappeared, and pollen tube growth resumed, supporting an essential role for apical actin in transporting secretory vesicles and pollen tube polar growth (Parton et al. 2001; De Graaf et al. 2005; Li et al. 2017).

In *Arabidopsis thaliana* root hairs, actin functions with similar characteristics as in pollen tubes. Fine actin filament bundles lie net-axially in cytoplasmic strands in the shank region, while actin bundles in subapical region are more diffused than in pollen tube and extend further into the apical region (Baluška et al. 2000; Hepler et al. 2001; Carol and Dolan 2002). Actin bundles in the root hair tube are reported to be involved in targeting and releasing the secretory vesicles to exocytosis sites, supporting the concept that transport of cellular components is needed for new plasma membrane and cell wall formation (Miller et al. 1999; Jones et al. 2002). Application of cytochalasin D (CD), known to stop actin filament elongation, affects Golgi vesicle targeting and inserting into the exocytosis site, and caused termination of root hair growth (Pollard and Mooseker 1981; Miller et al. 1999). Besides this, previous studies have shown that the depolymerization of F-actin bundles in vivo inhibited root hair growth (Li et al. 2018b). Both *actin2* and *actin8* knockdown mutants have shorter root hairs, while Actin2 was reported to be essential for root hair bulge site selection and tip growth (An et al. 1996; Ringli et al. 2002; Vaškebová et al. 2017). In the *Actin7* mutant, root hair density is reduced and more radial bulging was observed, suggesting that Actin7 participates in the early stage of root hair development (Kandasamy et al. 2001). In a recent study, Actin7 was shown to participate in auxin-mediated Rho-of-plant (ROP) signaling of polar epidermal root hair initiation (Kiefer et al. 2015). The planar root hair polarity in *actin7* mutant was disturbed (Kiefer et al. 2015; Zhu et al. 2016); however, this phenotype can be partially rescued by the synthetic auxin transport inhibitor, NPA (1-*N*-naphthylphthalamic acid) (Zhu et al. 2016).

Recently, genetic fusion of *actin2* and *actin7* with fluorescent proteins allowed observation of individual actin filaments. Surprisingly, in epidermal cells Actin2 was incorporated into thinner filaments, while Actin7 was associating with thicker bundles (Kijima et al. 2018).

7.2.3   Role of Actin-Binding Proteins (ABPs) in Polar Growth

Numerous actin-binding proteins (ABPs) have been described to modulate actin dynamics via affecting actin nucleation, elongation, depolymerization, severing, and bundling (Ren and Xiang 2007; Cheung and Wu 2008; Staiger et al. 2010; Fu 2015). Meanwhile, it has been established that ABP activities are regulated by several intracellular processes known to impact plant polar growth, such as local Ca^{2+} gradients, pH, and phosphoinositides (Franklin-Tong et al. 1996; Monteiro et al. 2005; Lovy-Wheeler et al. 2006; Saarikangas et al. 2010; Bezanilla et al. 2015).

Profilins

The actin monomer-binding proteins, profilins, regulate actin nucleation and enhance formin-mediated actin filament assembly at plus ends and can thereby regulate polar growth (Pollard and Borisy 2003; Suarez et al. 2015). Profilins are distributed throughout the pollen tube and functionally distinct as their action is affected by Ca^{2+} gradients (Kovar et al. 2000).

Arabidopsis thaliana encodes for three vegetative profilin isoforms (*PRF1, PRF2, PRF3*), and two reproductive isoforms (*PRF4, PRF5*) (Huang et al. 1996). Injecting profilin protein into pollen tubes severely disrupts actin organization and leads to shorter pollen tubes (Vidali et al. 2001), while in *prf4 prf5* mutants, the elongation rate of pollen tubes is significantly reduced (Liu et al. 2015). The vegetative *PRF1* plays an essential role in root hair growth (Baluška et al. 2000). Overexpression of *PRF1* results in longer root hairs and hypocotyls (Ramachandran et al. 2000). Reducing expression of *PRF1* by expressing a *PRF1* anti-sense construct leads to shorter root hairs, while *prf1* mutant plants reveal a high density of longer root hairs (Ramachandran et al. 2000; McKinney et al. 2001). Differences in these phenotypes are likely due to the high homology of the three vegetative profilins, which might allow that expression level of other profilins to be potentially also affected by the anti-sense probe of *PRF1* (Ramachandran et al. 2000; Pei et al. 2012).

Actin Depolymerizing Factors (ADFs)

Actin depolymerization factors (ADFs) bind both G-actin and F-actin, stimulate depolymerization of actin filaments at the minus end (Maciver and Hussey 2002), and also have bundling activity in the plant cell (Dong et al. 2001a). NtADF1 and AtADF7 are required for turnover of longitudinal actin cables by severing actin filaments in the shank region, inhibiting pollen tube growth (Chen et al. 2002; Zheng et al. 2013). Overexpressing of AtDF1 reduced longitudinal actin cables and resulted in shorter and thicker root hairs. In contrast, downregulation of *AtADF1* expression leads to longer root hairs via promotion of actin cable formation (Dong et al. 2001b). Actin-interacting protein, AIP1, enhances actin-depolymerizing activity in the presence of ADFs. Reduction of AIP1 results in the promotion of actin bundling and a short root hair phenotype (Ketelaar et al. 2004; Ono et al. 2004).

pH in the apical region of elongating pollen tubes differs, revealing an acidic tip (pH 6.0) and an alkaline subapical region (pH 7.6) (Feijó et al. 1999), and affects ADF activity. Slightly alkaline conditions promote the actin-depolymerizing activity of pollen ADFs in vitro, suggesting the possibility that the ADFs' activity is locally augmented in the subapical region (Allwood et al. 2002; Chen et al. 2002). Moreover, phosphatidylinositol 4,5-bisphosphate (PIP2), which is enriched in the apical membrane, can inhibit in vitro actin-binding activity of ADFs (Allwood et al. 2002; Dowd et al. 2006; Helling et al. 2006). However, the reduction of ADF activity by the acidic tip cytoplasm and the PIP2-enriched apical membrane seems to be important for nascent actin assembly along the apical and apical flank of membranes (Cheung and Wu 2008). Besides this, NtADF2 was considered being an important player in auxin-induced actin reorganization in BY-2 cell (Durst et al. 2013).

Villins

Members of the villin family display filament bundling, barbed-end capping, and severing activity and are responsive to calcium (Huang et al. 2005). The archetype member of the lily villin family, ABP29, retains a full suite of activities, including a severing activity in a Ca^{2+}- and/or phosphatidylinositol 4,5-bisphosphate-regulated manner in vitro. Transient overexpression of ABP29 in lily pollen resulted in actin filament fragmentation and inhibited pollen tube growth (Xiang et al. 2007). The reduction of ABP41 activity by injecting an ABP41 antibody into lily pollen leads to pollen tube growth inhibition (Fan et al. 2004).

In *Arabidopsis*, all of five villin isoforms (AtVLNs) are expressed ubiquitously and all except AtVLN2 bundle actin (Klahre et al. 2000). AtVLN4 is essential in root hair growth. In root hair of *atvln4* mutant, longitudinal actin cables are disrupted and thin and discontinuous short actin bundles exist in the shank region. As a result, cytoplasmic streaming is inhibited, resulting in shorter root hairs (Zhang et al. 2011a). AtVLN2 and AtVLN5 act redundantly to promote actin turnover in pollen tubes and facilitate the construction of the actin fringe (Zhang et al. 2010; Qu et al. 2013). In rice, mutation of *VILLIN2 (VLN2)* resulted in malformed organs, including twisted roots and shoots at the seeding stage, which is accompanied by a more dynamic actin cytoskeleton network than the wild type (Wu et al. 2015). Interestingly, the *vln2* mutant exhibits hypersensitive, gravitropic responses, faster recycling of PIN2, and altered auxin distribution (Wu et al. 2015). These results indicate that VLN2 plays an important role in regulating plant architecture by modulating actin dynamics and polar auxin transport.

Formins and the Actin-Related Protein 2/3 (Arp2/3) Complex

Formins and the actin-related protein 2/3 (Arp2/3) complex are actin nucleators that are both expressed in pollen. The Arp2/3 complex is made of seven-subunit proteins and capable of organizing filaments into branched networks by stimulating actin assembly with a fixed angle of 70° from the side of a preexisting actin filament (Mullins et al. 1998; Goley and Welch 2006; Pollard 2007). The Arp2/3 complex functions in cell motility, endocytosis, and membrane trafficking and is regulated by the key regulator for polar growth, so-called Rho-GTPases (Pellegrin and Mellor 2005; Ridley 2015). In maize root hairs, the subunit Arp3 of the Arp2/3 complex is localized along the cortical actin filaments and accumulates at the apical plasma membrane (Van Gestel et al. 2003). In *Arabidopsis*, knockout mutants of the Arp2/3 complex reveal wavy, swollen, and branched root hairs with reduced length (Mathur et al. 2003a, b). The Arp2/3 complex is involved in both root hair initiation and elongation; however, its role in pollen tubes is still unclear.

Formins activate de novo actin nucleation. They consist of multiple functional domains, including the highly conserved formin homology domain, FH1 and FH2 domains. Their elongation activity can be enhanced by interaction with profilins

(Romero et al. 2004; Paul and Pollard 2008; Blanchoin and Staiger 2010; Li et al. 2017).

In *Arabidopsis*, there are 21 formin isoforms that can be divided into two clades, class I and class II. All class I formins (except AtFH7) have an N-terminus with transmembrane domains, while class II formins, like AtFH13, AtFH14, or AtFH18, contain a phosphatase and a tensin-related (PTEN)-like domain in the N-terminus (Deeks et al. 2002). Overexpression of AtFH1, AtFH3, AtFH5, or AtFH8 induces morphological defects in tip-growing cells, such as pollen tubes or root hairs, by disrupting actin cables, which are essential tracks for the movement of large organelles and cytoplasmic streaming, or control membrane-originated actin polymerization at pollen tube tips (Cheung and Wu 2004; Yi et al. 2005; Ye et al. 2009; Cheung et al. 2010; Lan et al. 2018). A recent report showed that AtFH5 can even affect polarity establishment and vesicle trafficking during pollen germination (Liu et al. 2018a). In rice (*Oryza sativa*), Formin Homology1 (OsFH1) regulates root hair elongation (Huang et al. 2013). In the moss (*Physcomitrella patens*), the class II formin (PpFor2A) owning a PTEN domain localizes at the cell cortex by PI(3,5)P2 binding and mediates rapid actin filament elongation and is essential for the polarized tip cell growth of protonemata (Vidali et al. 2009b; van Gisbergen et al. 2012).

In animal and fungal formins, there is no evidence for an interaction between plant formins and ROPs; therefore, the activation mechanism of plant formins is still unclear (Deeks et al. 2002). The Rice Morphology Determinant (RMD), also known as Formin Homology5 (OsFH5), belongs to class II formins and localizes at the tip of the rice pollen tube. It affects apical actin density and longitudinal shank cable arrangements and is essential for cytoplasmic streaming during pollen tube growth (Yang et al. 2011; Zhang et al. 2011b; Li et al. 2018a). Further, it has been reported to be a crucial protein of the auxin–actin self-organizing regulatory loop (Li et al. 2014a), suggesting that the activation of plant formins might employ another mechanism (for details see below).

Myosins

Myosins are motor proteins owning ATPase and actin-binding activities and slide along filamentous actin (F-actin) (Sweeney and Houdusse 2010). Among the myosin classes, flowering plants harbor only myosin classes, VIII and XI (Odronitz and Kollmar 2007). In *Arabidopsis*, 4 myosin VIII and 13 myosin XI isoforms are present, which themselves are highly redundant (Reddy and Day 2001). Myosin XI (including XI-1, XI-2, XI-B, XI-I, and XI-K) is predominantly involved in organelle movement, cytoplasmic streaming (Peremyslov et al. 2008), and ER streaming by organizing actin bundles, especially myosin XI-K (Ueda et al. 2010, 2015; Griffing et al. 2014), and its activity is inhibited by high calcium conditions (Kohno and Shimmen 1988; Tominaga et al. 2012). Myosin XI affects polarized cell growth in root hairs and pollen tubes, and both myosin XI-2 and myosin XI-K mutant exhibited shorter root hair and reduced organelle trafficking (Ojangu et al. 2007; Peremyslov et al. 2008, 2010; Park and Nebenfuhr 2013). In pollen, six myosin XI isoforms (XI-A, XI-D, XI-B, XI-C, XI-E, XI-I, and XI-J) are highly expressed (Peremyslov et al. 2011; Sparkes 2011). In *XI-E XI-I* loss-of-function mutants, the F-actin is less bundled, more branched, and abnormally oriented in pollen tubes, and the organelle movement is reduced, resulting in a drastic inhibition of pollen tube growth (Madison et al. 2015). In tobacco, the RISAP (RAC5 interacting subapical pollen tube protein) directly interacts with myosin and F-actin (in a myosin-dependent manner) and is associated with the subapical TGN (Stephan et al. 2014); however, the detailed mechanism of how myosin XI affects F-actin organization remains unknown.

Fimbrins

Fimbrins are actin filament bundling protein, involved in the maintenance of longitudinal actin bundles in the shank region of pollen tubes. Loss of function of *AtFIM5* and *LiFIM1* results in actin filament disorganization in the shank, subapical, and apical region of pollen tube, Loss of function of AtFIM5 also exhibited disruption of cytoplasmic streaming, resulting in pollen tubes and root hairs of *Arabidopsis thaliana* with shorter length or larger width (Wu et al. 2010b; Su et al. 2012a; Zhang et al. 2016). The activity of LiFIM1 is sensitive to pH changes (Su et al. 2012a). Furthermore, AtFIM4 and AtFIM5 act synergistically during polarized pollen tube growth, also shown to regulate the growth of root hairs in an auxin-independent way, suggesting that AtFIM4 and AtFIM5 may be involved in auxin signaling pathways during root hair growth (Su et al. 2017; Ding et al. 2018a).

7.3 Auxin–Actin Cross Talk in Polar Growth

The asymmetric distribution of auxin (auxin gradients) is provided by polar auxin transport (PAT) and plays a major role in cell polarity establishment, such as planar root hair polarity (Geisler et al. 2014; Kiefer et al. 2015). Moreover, auxin was also shown to promote plant polar growth in root hairs and pollen tubes (Grones and Friml 2015). Polar auxin transport is tightly connected with the actin cytoskeleton because auxin transporters reach their final destination by trafficking of myosin-mediated vesicles that move along actin cables (Kleine-Vehn et al. 2008). Auxin transporter polarization and auxin efflux are thought to form a feedback loop that affects actin cytoskeleton reorganization and finally affects auxin polar transport. The auxin–actin cross talk provides thus a potential mechanism for auxin to control plant polar growth.

7.3.1 Self-Regulation of Polar Auxin Transport

It is known that auxin has the ability to regulate its own transport by distinct transcriptional and non-transcriptional pathways (Xu et al. 2014; Grones and Friml 2015). Here we briefly summarize the self-regulation of polar auxin transport and the involvement of actin.

Transcriptional Feedback Model

The best characterized auxin-dependent transcription perception system in the nucleus includes auxin co-receptor composed of transport inhibitor 1 (TIR1) or auxin signaling F-box (AFB) and auxin- or indole-3-acetic-acid-inducible family (AUX/IAA) (Ruegger et al. 1998; Dharmasiri et al. 2005a; Kepinski and Leyser 2005; Tan et al. 2007; Villalobos et al. 2012; Grones and Friml 2015). TIR1 and AFB are F-box proteins which are part of the E3 ubiquitin ligase complex $SCF^{TIR1/AFB}$. At low auxin concentrations, the IAA repressors, AUX/IAA, are associated with ARF proteins, functioning as transcription factors, and thus inhibit auxin-induced gene expression (Guilfoyle and Hagen 2007). High concentration of auxin activates the $SCF^{TIR1/AFB}$ complex by targeted ubiquitination of AUX/IAA proteins leading to their destruction via 26S proteasome-mediated degradation (Gray et al. 2001; Dharmasiri et al. 2005b; Kepinski and Leyser 2005; Petroski and Deshaies 2005; Tan et al. 2007; dos Santos Maraschin et al. 2009; Grones and Friml 2015). ARFs released from the AUX/IAA–ARF heteromer form ARF–ARF dimers that induce the expression (Guilfoyle and Hagen 2007; Korasick et al. 2014; Nanao et al. 2014).

Non-Transcriptional Feedback Model

High cellular auxin is thought to inhibit PIN1 endocytosis via the activation of the ROP2-RIC4 signaling pathway. In *Arabidopsis* in leaf pavement cells, auxin activates both ROP2-RIC4 and ROP6-RIC1 pairs (Xu et al. 2010). ROP2-RIC4 functions to reduce PIN1 endocytosis and promote the lobe plasma membrane location of PIN1, via the accumulation of cortical actin microfilaments induced by the POP2 effector protein RIC4 (ROP-interactive CRIB-containing proteins, RIC) (Fu et al. 2002). By contrast, ROP6-RIC1 functions on microtubules and inhibits exocytosis (Fu et al. 2005). Recent reports show that ROP6-RIC1 also regulates the association of clathrin with the plasma membrane for clathrin-mediated endocytosis (Chen et al. 2012).

RIC1 is also expressed in pollen tubes. In *RIC1* loss-of-function mutants, the length of the pollen tubes is enhanced, while *RIC1* overexpression leads to a dramatically reduced length of the pollen tubes. Unexpectedly, RIC1 prefers severing actin in a Ca^{2+}-dependent fashion rather than regulating microtubules, suggesting a distinct function for RIC1 in pollen (Zhou et al. 2015).

For years, the impact of auxin on the ROP-RIC pathway mediating the rapid, non-transcriptional asymmetrical distribution of PIN1 in a SCF$^{TIR1/AFB}$-independent way was thought to be initialized by Auxin-Binding Protein1 (ABP1) functioning as an apoplastic auxin receptor (Rück et al. 1993; Steffens et al. 2001; Paciorek et al. 2005; Robert et al. 2010; Xu et al. 2010, 2014). As part of a cell surface auxin-sensing complex, ABP1 was thought to activate ROP-RIC signaling by interacting with transmembrane kinase (TMK) receptor-like kinases in an auxin-dependent fashion that would transduce this information to the cytoplasm (Xu et al. 2014).

However, ABP1 apparently is not part of this signaling cascade (Di et al. 2015; Paponov et al. 2018), and the embryo-lethal phenotypes of *abp1-1* alleles are most likely caused by second-site mutation (Dai et al. 2015; Michalko et al. 2015). Newer work indicated that auxin transport-mediated hyperpolarization itself could be the driving force for altered PIN endocytosis (Paponov et al. 2018).

Over the last few years, the FKBP42, Twisted Dwarf1 (TWD1), was established both as a relevant regulator and as a co-chaperone of auxin-transporting ABCB (Geisler et al. 2003, 2005; Wu et al. 2010a; Wang et al. 2013; Di Donato 2017). Interestingly, TWD1 was shown to physically interact with Actin7, although likely indirect, and to regulate actin filament organization and dynamics (Zhu et al. 2016). Also, TWD1 is required for NPA-mediated actin cytoskeleton remodeling. It appears that TWD1 determines downstream locations of auxin efflux transporters by adjusting actin filament de-bundling and dynamizing processes and mediates NPA action on the latter. As a consequence, *actin7* and *twd1* share developmental and physiological phenotypes indicative of defects in auxin transport, which can be phenocopied by NPA treatment or by chemical actin (de)stabilization.

7.3.2 Auxin Affects Actin Expression and Reorganization

As mentioned above, there exist three vegetative actin isoforms in *Arabidopsis* and all of them are involved in root hair development. *Actin7*, the only isoform that responds strongly to exogenous auxin, is essential for hormone-stimulated callus formation (McDowell et al. 1996). Actin7 can interact with AIP1-2 and is involved in auxin-mediated polar recruitment of ROP to sites of polar epidermal hair initiation (Kiefer et al. 2015). Interestingly, blocking PAT (polar auxin transport) by using the PAT inhibitor, NPA, resulted in similar phenotypic defects as loss of *Actin7* (Zhu et al. 2016).

Recent reports show that the ER-localized PIN proteins, PIN5 and PIN8, are specifically expressed in pollen tubes. Overexpression of *PIN8* enhances pollen tube elongation and increases the resistance of pollen to the auxin efflux inhibitor, NPA (Ding et al. 2012). In *pin8* and *pin5* loss-of-function mutants, the germination rate of pollen grain is reduced, which is of interest because the germination process is considered to be dependent on the actin turnover (Bosco et al. 2012; Ding et al. 2012; Cao et al. 2013; Liu et al. 2018b). However, the mechanism of how PIN5 and PIN8 regulate pollen tube elongation is still unclear.

7.3.3 Auxin Affects Actin-Binding Protein Expression

Auxin affects not only expression of actin but also expression of actin-binding protein. An exciting example is RMD, a classic type of formin in rice, containing a PTEN-like domain. *rmd* mutant shows reduced sensitivity to IAA, shorter root, and smaller pollen (Yang et al. 2011; Zhang et al. 2011b; Li et al. 2018a). Reduced IAA sensitivity of root growth corresponds to the finding that IAA (10 μM, 6 h) induces F-actin bundling in root cells, which is absent in *rmd* mutants, indicating that RMD is essential for the auxin-mediated reorganization of F-actin arrays (Yang et al. 2011; Zhang et al. 2011b). Now it is clear that the expression of RMD is affected by the auxin response factors, OsARF23 and OsARF24, heterodimers (Li et al. 2014a). Auxin activates the activity of OsARF23 and OsARF24, which together bind to promoter region and induce the expression of the *rmd* gene (Li et al. 2014a).

The *rmd* mutant exhibits abnormal pollen tube growth and a decreased pollen germination rate both in vitro and in vivo. The F-actin polarity and distribution are disrupted in the pollen tube (Li et al. 2018a). This suggests that the tip-localized RMD is essential for pollen tube growth and polarity as well as F-actin organization, while the tip localization of RMD relies on the PTEN-like domain (Li et al. 2018a). Recently, it was also shown that RMD controls root growth angles by linking actin filaments and the gravity sensing organelles, statoliths. RMD was localized to the surface of statoliths, while *rmd* mutants reveal faster root bending, most likely due to more rapid statolith movements (Huang et al. 2018). This unexpected phenotype again supports a tight link between actin filaments and root gravitropsim.

7.4 Conclusion and Outlook

In this chapter, we have gathered information on how auxin and actin interfere during the process of plant polar growth, using pollen tubes and root hairs as an example. A review of older and more recent literature indicates that auxin and actin indeed cooperate on the transcriptional but also on the posttranscriptional level as summarized in Fig. 7.1. The upregulation of *ABPs* (like RMD; Li et al. 2014a) and actin isoforms (like ACTIN7; McDowell et al. 1996) itself as well as that of auxin transporters from different subclasses (like some *PIN* and *ABCB* isoforms) is well known for some time. Likewise, it is clear for quite a while that auxin transporter polarization and auxin efflux form a feedback loop that not only involves but also affects actin cytoskeleton reorganization, and as a consequence thus finally also polar auxin transport (Maisch and Nick 2007; Rahman et al. 2007; Nick et al. 2009). The collected data now indicate that an auxin–actin cross talk provides a very likely venue for auxin to control plant polar growth by influencing actin dynamics. However, while during leaf development, local auxin maxima are thought to be

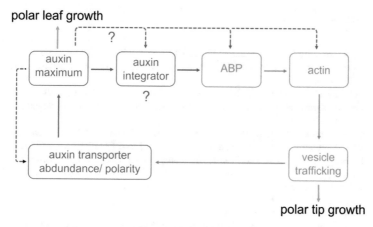

Fig. 7.1 Hypothetical model on the cooperative action of auxin and actin during polar growth. The collected evidence indicates that auxin and related elements (in blue) and actin-related elements (in orange) cooperate on the transcriptional (solid lines) but also on the posttranscriptional level (dashed lines). While during leaf development, local auxin maxima are thought to be the primary driving force for polar outgrowth, polar growth in pollen tubes and root hairs seems to employ vesicle trafficking as primary spearhead. Please note that so-called auxin integrators (Zhu and Geisler 2015) bridging auxin itself and the functionality of different subclasses of ABPs have not been identified; also it is unclear if auxin has an effect on the integrator expression

the primary driving force for polar outgrowth (Sandalio et al. 2016), polar growth in pollen tubes and root hairs seems to employ vesicle trafficking as primary spearhead (Fig. 7.1; Campanoni and Blatt 2006; Cole and Fowler 2006). Obviously, these two driving forces for polar growth are tightly connected and by nature hard to separate in such a kind of self-amplifying loop.

While many components of this regulatory feedback loop have been identified, it is still somewhat surprising that the central elements that would function as a bridge between auxin itself and the functionality of different subclasses of ABPs, so-called auxin integrators (Zhu et al. 2016), have not been identified. After the evaporation of ABP1 as the primary candidate (Xu et al. 2010), the identification of such integrating factors should have the highest priority. A plausible candidate might be in fact among others, the FKBP42, TWD1, was shown to bind (indirectly) and to regulate actin dynamics and also to be responsible for auxin transporter trafficking. Currently, it is however still unclear if these two functionalities on TWD1 are connected and if TWD1 binds indeed auxin, which would be a primary criterion.

Acknowledgments   The author would like to thank J. Zhu for discussion and comments. Work in our lab is currently funded by the Swiss National Funds (project 31003A_165877/1) and the European Space Association (CORA-GBF project LIRAT).

References

Allwood EG, Anthony RG, Smertenko AP, Reichelt S, Drobak BK, Doonan JH et al (2002) Regulation of the pollen-specific actin-depolymerizing factor LlADF1. Plant Cell 14(11):2915–2927

An YQ, McDowell JM, Huang S, McKinney EC, Chambliss S, Meagher RB (1996) Strong, constitutive expression of the Arabidopsis ACT2/ACT8 actin subclass in vegetative tissues. Plant J 10(1):107–121

Andersland JM, Fisher DD, Wymer CL, Cyr RJ, Parthasarathy MV (1994) Characterization of a monoclonal antibody prepared against plant actin. Cell Motil Cytoskeleton 29(4):339–344

Baluška F, Salaj J, Mathur J, Braun M, Jasper F, Šamaj J et al (2000) Root hair formation: F-actin-dependent tip growth is initiated by local assembly of profilin-supported F-actin meshworks accumulated within expansin-enriched bulges. Dev Biol 227(2):618–632

Bezanilla M, Gladfelter AS, Kovar DR, Lee W-L (2015) Cytoskeletal dynamics: a view from the membrane. J Cell Biol 209(3):329–337

Bibikova TN, Blancaflor EB, Gilroy S (1999) Microtubules regulate tip growth and orientation in root hairs of *Arabidopsis thaliana*. Plant J 17(6):657–665

Blanchoin L, Staiger CJ (2010) Plant formins: diverse isoforms and unique molecular mechanism. Biochim Biophys Acta 1803(2):201–206

Boavida LC, McCormick S (2007) TECHNICAL ADVANCE: temperature as a determinant factor for increased and reproducible in vitro pollen germination in *Arabidopsis thaliana*. Plant J 52(3):570–582

Bosco CD, Dovzhenko A, Liu X, Woerner N, Rensch T, Eismann M et al (2012) The endoplasmic reticulum localized PIN8 is a pollen-specific auxin carrier involved in intracellular auxin homeostasis. Plant J 71(5):860–870

Campanoni P, Blatt MR (2006) Membrane trafficking and polar growth in root hairs and pollen tubes. J Exp Bot 58(1):65–74

Cao L-J, Zhao M-M, Liu C, Dong H-J, Li W-C, Ren H-Y (2013) LlSR28 is involved in pollen germination by affecting filamentous actin dynamics. Mol Plant 6(4):1163–1175

Carol RJ, Dolan L (2002) Building a hair: tip growth in *Arabidopsis thaliana* root hairs. Philos Trans R Soc Lond B Biol Sci 357(1422):815–821

Cecchetti V, Altamura MM, Falasca G, Costantino P, Cardarelli M (2008) Auxin regulates Arabidopsis anther dehiscence, pollen maturation, and filament elongation. Plant Cell 20(7):1760–1774

Chen D, Zhao J (2008) Free IAA in stigmas and styles during pollen germination and pollen tube growth of *Nicotiana tabacum*. Physiol Plant 134(1):202–215

Chen CY, Wong EI, Vidali L, Estavillo A, Hepler PK, Wu H-M et al (2002) The regulation of actin organization by actin-depolymerizing factor in elongating pollen tubes. Plant Cell 14(9):2175–2190

Chen X, Naramoto S, Robert S, Tejos R, Lofke C, Lin D et al (2012) ABP1 and ROP6 GTPase signaling regulate clathrin-mediated endocytosis in Arabidopsis roots. Curr Biol 22(14):1326–1332. https://doi.org/10.1016/j.cub.2012.05.020

Cheung AY, Wu H-M (2004) Overexpression of an Arabidopsis formin stimulates supernumerary actin cable formation from pollen tube cell membrane. Plant Cell 16(1):257–269

Cheung AY, Wu H-M (2007) Structural and functional compartmentalization in pollen tubes. J Exp Bot 58(1):75–82. https://doi.org/10.1093/jxb/erl122

Cheung AY, Wu H-M (2008) Structural and signaling networks for the polar cell growth machinery in pollen tubes. Annu Rev Plant Biol 59:547–572

Cheung AY, Niroomand S, Zou Y, Wu H-M (2010) A transmembrane formin nucleates subapical actin assembly and controls tip-focused growth in pollen tubes. Proc Natl Acad Sci USA 107(37):16390–16395

Cole RA, Fowler JE (2006) Polarized growth: maintaining focus on the tip. Curr Opin Plant Biol 9 (6):579–588

Dai X, Zhang Y, Zhang D, Chen J, Gao X, Estelle M et al (2015) Embryonic lethality of Arabidopsis abp1-1 is caused by deletion of the adjacent BSM gene. Nat Plants 1:15183. https://doi.org/10.1038/nplants.2015.183

De Graaf BH, Cheung AY, Andreyeva T, Levasseur K, Kieliszewski M, Wu H-M (2005) Rab11 GTPase-regulated membrane trafficking is crucial for tip-focused pollen tube growth in tobacco. Plant Cell 17(9):2564–2579

Deeks MJ, Hussey PJ, Davies B (2002) Formins: intermediates in signal-transduction cascades that affect cytoskeletal reorganization. Trends Plant Sci 7(11):492–498

Del Casino C, Li YQ, Moscatelli A, Scali M, Tiezzi A, Cresti M (1993) Distribution of microtubules during the growth of tobacco pollen tubes. Biol Cell 79(2):125–132

Derksen J, Rutten T, Lichtscheidl I, De Win A, Pierson E, Rongen G (1995) Quantitative analysis of the distribution of organelles in tobacco pollen tubes: implications for exocytosis and endocytosis. Protoplasma 188(3–4):267–276

Dharmasiri N, Dharmasiri S, Estelle M (2005a) The F-box protein TIR1 is an auxin receptor. Nature 435:441. https://doi.org/10.1038/nature03543

Dharmasiri N, Dharmasiri S, Weijers D, Lechner E, Yamada M, Hobbie L et al (2005b) Plant development is regulated by a family of auxin receptor F box. Proteins 9(1):109–119

Di D-W, Zhang C, Guo G-Q (2015) Involvement of secondary messengers and small organic molecules in auxin perception and signaling. Plant Cell Rep 34(6):895–904

Di Donato M (2017) Regulation of ABCB-mediated auxin transport by HSP90 and TWISTED DWARF1

Ding Z, Wang B, Moreno I, Dupláková N, Simon S, Carraro N et al (2012) ER-localized auxin transporter PIN8 regulates auxin homeostasis and male gametophyte development in Arabidopsis. Nat Commun 3:941

Ding X, Zhang S, Liu J, Liu S, Su H (2018a) Arabidopsis FIM4 and FIM5 regulates the growth of root hairs in an auxin-insensitive way. Plant Signal Behav 13(9):e1473667. https://doi.org/10.1080/15592324.2018.1473667

Ding X, Zhang S, Liu J, Liu S, Su H (2018b) Arabidopsis FIM4 and FIM5 regulates the growth of root hairs in an auxin-insensitive way. Plant Signal Behav 13(9):e1473667

Dong C-H, Kost B, Xia G, Chua N-H (2001a) Molecular identification and characterization of the Arabidopsis AtADF1, AtADF5 and AtADF6 genes. Plant Mol Biol 45(5):517–527

Dong C-H, Xia G-X, Hong Y, Ramachandran S, Kost B, Chua N-H (2001b) ADF proteins are involved in the control of flowering and regulate F-actin organization, cell expansion, and organ growth in Arabidopsis. Plant Cell 13(6):1333–1346

dos Santos Maraschin F, Memelink J, Offringa R (2009) Auxin-induced, SCFTIR1-mediated polyubiquitination marks AUX/IAA proteins for degradation. Plant J 59(1):100–109

Dowd PE, Coursol S, Skirpan AL, Kao T-H, Gilroy S (2006) Petunia phospholipase C1 is involved in pollen tube growth. Plant Cell 18(6):1438–1453

Duan Z, Tominaga M (2018) Actin–myosin XI: an intracellular control network in plants. Biochem Biophys Res Commun 506(2):403–408

Durst S, Nick P, Maisch J (2013) Nicotiana tabacum actin-depolymerizing factor 2 is involved in actin-driven, auxin-dependent patterning. J Plant Physiol 170(12):1057–1066

Fan X, Hou J, Chen X, Chaudhry F, Staiger CJ, Ren H (2004) Identification and characterization of a Ca2+-dependent actin filament-severing protein from lily pollen. Plant Physiol 136 (4):3979–3989

Feijó J, Sainhas J, Hackett G, Kunkel J, Hepler P (1999) Growing pollen tubes possess a constitutive alkaline band in the clear zone and a growth-dependent acidic tip. J Cell Biol 144 (3):483–496

Feijó JA, Costa SS, Prado AM, Becker JD, Certal AC (2004) Signalling by tips. Curr Opin Plant Biol 7(5):589–598

Feng X-L, Ni W-M, Elge S, Mueller-Roeber B, Xu Z-H, Xue H-W (2006) Auxin flow in anther filaments is critical for pollen grain development through regulating pollen mitosis. Plant Mol Biol 61(12):215–226

Foissner I, Grolig F, Obermeyer G (2002) Reversible protein phosphorylation regulates the dynamic organization of the pollen tube cytoskeleton: effects of calyculin A and okadaic acid. Protoplasma 220(1–2):0001–0015

Franklin-Tong VE, Drobak BK, Allan AC, Watkins PA, Trewavas AJ (1996) Growth of pollen tubes of *Papaver rhoeas* is regulated by a slow-moving calcium wave propagated by inositol 1, 4, 5-trisphosphate. Plant Cell 8(8):1305–1321

Fu Y (2015) The cytoskeleton in the pollen tube. Curr Opin Plant Biol 28:111–119

Fu Y, Wu G, Yang Z (2001) Rop GTPase-dependent dynamics of tip-localized F-actin controls tip growth in pollen tubes. J Cell Biol 152(5):1019–1032

Fu Y, Li H, Yang Z (2002) The ROP2 GTPase controls the formation of cortical fine F-actin and the early phase of directional cell expansion during Arabidopsis. Organogenesis 14(4):777–794

Fu Y, Gu Y, Zheng Z, Wasteneys G, Yang Z (2005) Arabidopsis interdigitating cell growth requires two antagonistic pathways with opposing action on cell morphogenesis. Cell 120(5):687–700

Galway ME, Heckman JW Jr, Schiefelbein JW (1997) Growth and ultrastructure of Arabidopsis root hairs: the rhd3 mutation alters vacuole enlargement and tip growth. Planta 201(2):209–218. https://doi.org/10.1007/bf01007706

Geisler M, Kolukisaoglu HU, Bouchard R, Billion K, Berger J, Saal B et al (2003) TWISTED DWARF1, a unique plasma membrane-anchored immunophilin-like protein, interacts with Arabidopsis multidrug resistance-like transporters AtPGP1 and AtPGP19. Mol Biol Cell 14(10):4238–4249

Geisler M, Blakeslee JJ, Bouchard R, Lee OR, Vincenzetti V, Bandyopadhyay A et al (2005) Cellular efflux of auxin catalyzed by the Arabidopsis MDR/PGP transporter AtPGP1. Plant J 44(2):179–194

Geisler M, Wang B, Zhu J (2014) Auxin transport during root gravitropism: transporters and techniques. Plant Biol (Stuttg) 16:50–57

Geitmann A, Emons AM (2000) The cytoskeleton in plant and fungal cell tip growth. J Microsc 198(3):218–245

Gilroy S, Jones DL (2000) Through form to function: root hair development and nutrient uptake. Trends Plant Sci 5(2):56–60

Goley ED, Welch MD (2006) The ARP2/3 complex: an actin nucleator comes of age. Nat Rev Mol Cell Biol 7(10):713

Gomez TM, Spitzer NC (1999) In vivo regulation of axon extension and pathfinding by growth-cone calcium transients. Nature 397(6717):350

Gossot O, Geitmann A (2007) Pollen tube growth: coping with mechanical obstacles involves the. Cytoskeleton 226(2):405–416

Gray WM, Kepinski S, Rouse D, Leyser O, Estelle M (2001) Auxin regulates SCF TIR1-dependent degradation of AUX/IAA. Proteins 414(6861):271

Griffing LR, Gao HT, Sparkes I (2014) ER network dynamics are differentially controlled by myosins XI-K, XI-C, XI-E, XI-I, XI-1, and XI-2. Front Plant Sci 5:218. https://doi.org/10.3389/fpls.2014.00218

Grones P, Friml J (2015) Auxin transporters and binding proteins at a glance. J Cell Sci 128(1):1–7

Guilfoyle TJ, Hagen G (2007) Auxin response factors. Curr Opin Plant Biol 10(5):453–460

Helling D, Possart A, Cottier S, Klahre U, Kost B (2006) Pollen tube tip growth depends on plasma membrane polarization mediated by tobacco PLC3 activity and endocytic membrane recycling. Plant Cell 18(12):3519–3534

Hepler PK, Winship LJ (2015) The pollen tube clear zone: clues to the mechanism of polarized growth. J Integr Plant Biol 57(1):79–92

Hepler PK, Vidali L, Cheung AY (2001) Polarized cell growth in higher plants. Annu Rev Cell Dev Biol 17(1):159–187

Huang S, McDowell JM, Weise MJ, Meagher RB (1996) The Arabidopsis profilin gene family (evidence for an ancient split between constitutive and pollen-specific profilin genes). Plant Physiol 111(1):115–126

Huang S, Robinson RC, Gao LY, Matsumoto T, Brunet A, Blanchoin L et al (2005) Arabidopsis VILLIN1 generates actin filament cables that are resistant to depolymerization. Plant Cell 17(2):486–501

Huang J, Kim CM, Xuan YH, Liu J, Kim TH, Kim BK et al (2013) Formin homology 1 (OsFH1) regulates root-hair elongation in rice (*Oryza sativa*). Planta 237(5):1227–1239. https://doi.org/10.1007/s00425-013-1838-8

Huang G, Liang W, Sturrock CJ, Pandey BK, Giri J, Mairhofer S et al (2018) Rice actin binding protein RMD controls crown root angle in response to external phosphate. Nat Commun 9(1):2346

Jones MA, Shen J-J, Fu Y, Li H, Yang Z, Grierson CS (2002) The Arabidopsis Rop2 GTPase is a positive regulator of both root hair initiation and tip growth. Plant Cell 14(4):763–776

Jones AR, Kramer EM, Knox K, Swarup R, Bennett MJ, Lazarus CM et al (2009) Auxin transport through non-hair cells sustains root-hair. Development 11(1):78

Kandasamy MK, Gilliland LU, McKinney EC, Meagher RB (2001) One plant actin isovariant, ACT7, is induced by auxin and required for normal callus formation. Plant Cell 13 (7):1541–1554

Kepinski S, Leyser O (2005) The Arabidopsis F-box protein TIR1 is an auxin receptor. Nature 435:446. https://doi.org/10.1038/nature03542

Ketelaar T, Allwood EG, Anthony R, Voigt B, Menzel D, Hussey PJ (2004) The actin-interacting protein AIP1 is essential for actin organization and plant. Development 14(2):145–149

Kiefer CS, Claes AR, Nzayisenga J-C, Pietra S, Stanislas T, Hüser A et al (2015) Arabidopsis AIP1-2 restricted by WER-mediated patterning modulates planar polarity. Development 142 (1):151–161

Kijima ST, Staiger CJ, Katoh K, Nagasaki A, Ito K, Uyeda TQP (2018) Arabidopsis vegetative actin isoforms, AtACT2 and AtACT7, generate distinct filament arrays in living plant cells. Sci Rep 8(1):4381. https://doi.org/10.1038/s41598-018-22707-w

Klahre U, Friederich E, Kost B, Louvard D, Chua N-H (2000) Villin-like actin-binding proteins are expressed ubiquitously in Arabidopsis. Plant Physiol 122(1):35–48

Kleine-Vehn J, Friml J (2008) Polar targeting and endocytic recycling in auxin-dependent plant development. Annu Rev Cell Dev Biol 24:447–473

Kohno T, Shimmen T (1988) Accelerated sliding of pollen tube organelles along Characeae actin bundles regulated by Ca2+. J Cell Biol 106(5):1539–1543

Koltai H, Dor E, Hershenhorn J, Joel DM, Weininger S, Lekalla S et al (2010) Strigolactones' effect on root growth and root-hair elongation may be mediated by auxin-efflux carriers. J Plant Growth Regul 29(2):129–136

Korasick DA, Westfall CS, Lee SG, Nanao MH, Dumas R, Hagen G et al (2014) Molecular basis for AUXIN RESPONSE FACTOR protein interaction and the control of auxin response repression. Proc Natl Acad Sci USA 111(14):5427–5432

Kovar DR, Drøbak BK, Staiger CJ (2000) Maize profilin isoforms are functionally distinct. Plant Cell 12(4):583–598

Kroeger JH, Daher FB, Grant M, Geitmann A (2009) Microfilament orientation constrains vesicle flow and spatial distribution in growing pollen tubes. Biophys J 97(7):1822–1831

Laitiainen E, Nieminen KM, Vihinen H, Raudaskoski M (2002) Movement of generative cell and vegetative nucleus in tobacco pollen tubes is dependent on microtubule cytoskeleton but independent of the synthesis of callose plugs. Sex Plant Reprod 15(4):195–204

Lan Y, Liu X, Fu Y, Huang S (2018) Arabidopsis class I formins control membrane-originated actin polymerization at pollen tube tips. PLoS Genet 14(11):e1007789. https://doi.org/10.1371/journal.pgen.1007789

Li G, Liang W, Zhang X, Ren H, Hu J, Bennett MJ et al (2014a) Rice actin-binding protein RMD is a key link in the auxin-actin regulatory loop that controls cell growth. Proc Natl Acad Sci USA 111(28):10377–10382. https://doi.org/10.1073/pnas.1401680111

Li G, Liang W, Zhang X, Ren H, Hu J, Bennett MJ et al (2014b) Rice actin-binding protein RMD is a key link in the auxin–actin regulatory loop that controls cell growth. Proc Natl Acad Sci USA 111(28):10377–10382. https://doi.org/10.1073/pnas.1401680111

Li J, Blanchoin L, Staiger CJ (2015) Signaling to actin stochastic dynamics. Annu Rev Plant Biol 66:415–440

Li S, Dong H, Pei W, Liu C, Zhang S, Sun T et al (2017) Ll FH 1-mediated interaction between actin fringe and exocytic vesicles is involved in pollen tube tip growth. New Phytol 214 (2):745–761

Li G, Yang X, Zhang X, Song Y, Liang W, Zhang D (2018a) Rice morphology determinant-mediated actin filament organization contributes to pollen tube growth. Plant Physiol 177 (1):255–270. https://doi.org/10.1104/pp.17.01759

Li J, Chen S, Wang X, Shi C, Liu H, Yang J et al (2018b) Hydrogen sulfide disturbs actin polymerization via S-sulfhydration resulting in stunted root hair growth. Plant Physiol 178 (2):936–949

Liu X, Qu X, Jiang Y, Chang M, Zhang R, Wu Y et al (2015) Profilin regulates apical actin polymerization to control polarized pollen tube growth. Mol Plant 8(12):1694–1709

Liu C, Zhang Y, Ren H (2018a) Actin polymerization mediated by AtFH5 directs the polarity establishment and vesicle trafficking for pollen germination in Arabidopsis. Mol Plant 11 (11):1389–1399. https://doi.org/10.1016/j.molp.2018.09.004

Liu C, Zhang Y, Ren H (2018b) Actin polymerization mediated by AtFH5 directs the polarity establishment and vesicle trafficking for pollen germination in Arabidopsis. Mol Plant 11 (11):1389–1399

Lord E (2000) Adhesion and cell movement during pollination: cherchez la femme. Trends Plant Sci 5(9):368–373

Lovy-Wheeler A, Wilsen KL, Baskin TI, Hepler PK (2005) Enhanced fixation reveals the apical cortical fringe of actin filaments as a consistent feature of the pollen tube. Planta 221(1):95–104

Lovy-Wheeler A, Kunkel JG, Allwood EG, Hussey PJ, Hepler PK (2006) Oscillatory increases in alkalinity anticipate growth and may regulate actin dynamics in pollen tubes of lily. Plant Cell 18(9):2182–2193

Maciver SK, Hussey PJ (2002) The ADF/cofilin family: actin-remodeling proteins. Genome Biol 3 (5):reviews3007. 3001

Madison SL, Buchanan ML, Glass JD, McClain TF, Park E, Nebenführ A (2015) Class XI myosins move specific organelles in pollen tubes and are required for normal fertility and pollen tube growth in Arabidopsis. Plant Physiol 169(3):1946–1960. https://doi.org/10.1104/pp.15.01161

Maisch J, Nick P (2007) Actin is involved in auxin-dependent patterning. Plant Physiol 143 (4):1695–1704

Mathur J, Mathur N, Kernebeck B, Hülskamp M (2003a) Mutations in actin-related proteins 2 and 3 affect cell shape development in Arabidopsis. Plant Cell 15(7):1632–1645

Mathur J, Mathur N, Kirik V, Kernebeck B, Srinivas BP, Hülskamp M (2003b) Arabidopsis CROOKED encodes for the smallest subunit of the ARP2/3 complex and controls cell shape by region specific fine F-actin formation. Development 130(14):3137–3146

McDowell JM, An Y, Huang S, McKinney EC, Meagher RB (1996) The Arabidopsis ACT7 actin gene is expressed in rapidly developing tissues and responds to several external stimuli. Plant Physiol 111(3):699–711

McKinney EC, Kandasamy MK, Meagher RB (2001) Small changes in the regulation of one Arabidopsis profilin isovariant, PRF1, alter seedling development. Plant Cell 13(5):1179–1191

Messerli M, Robinson KR (1997) Tip localized Ca2+ pulses are coincident with peak pulsatile growth rates in pollen tubes of *Lilium longiflorum*. J Cell Sci 110(11):1269–1278

Michalko J, Dravecka M, Bollenbach T, Friml J (2015) Embryo-lethal phenotypes in early abp1 mutants are due to disruption of the neighboring BSM gene. F1000Res 4:1104. https://doi.org/10.12688/f1000research.7143.1

Miller DD, De Ruijter NC, Bisseling T, Emons AMC (1999) The role of actin in root hair morphogenesis: studies with lipochito-oligosaccharide as a growth stimulator and cytochalasin as an actin perturbing drug. Plant J 17(2):141–154

Monteiro D, Coelho PC, Rodrigues C, Camacho L, Quader H, Malho R (2005) Modulation of endocytosis in pollen tube growth by phosphoinositides and phospholipids. Protoplasma 226 (1–2):31–38

Mullins RD, Heuser JA, Pollard TD (1998) The interaction of Arp2/3 complex with actin: nucleation, high affinity pointed end capping, and formation of branching networks of filaments. Proc Natl Acad Sci USA 95(11):6181–6186

Nagawa S, Xu T, Lin D, Dhonukshe P, Zhang X, Friml J et al (2012) ROP GTPase-dependent actin microfilaments promote PIN1 polarization by localized inhibition of clathrin-dependent endo-cytosis. PLoS Biol 10(4):e1001299

Nanao MH, Vinos-Poyo T, Brunoud G, Thévenon E, Mazzoleni M, Mast D et al (2014) Structural basis for oligomerization of auxin transcriptional regulators. Nat Commun 5:3617 (Supplementary information). https://doi.org/10.1038/ncomms4617. https://www.nature.com/articles/ncomms4617

Nebenführ A, Dixit R (2018) Kinesins and myosins: molecular motors that coordinate cellular functions in plants. Annu Rev Plant Biol 69:329–361

Nick P, Han M-J, An G (2009) Auxin stimulates its own transport by shaping actin filaments. Plant Physiol 151(1):155–167

Odronitz F, Kollmar M (2007) Drawing the tree of eukaryotic life based on the analysis of 2,269 manually annotated myosins from 328 species. Genome Biol 8(9):R196

Ojangu E-L, Järve K, Paves H, Truve E (2007) Arabidopsis thaliana myosin XIK is involved in root hair as well as trichome morphogenesis on stems and leaves. Protoplasma 230 (3–4):193–202

Ono S, Mohri K, Ono K (2004) Microscopic evidence that actin-interacting protein 1 actively disassembles actin-depolymerizing factor/cofilin-bound actin filaments. J Biol Chem 279 (14):14207–14212

Paciorek T, Zažímalová E, Ruthardt N, Petrášek J, Stierhof Y-D, Kleine-Vehn J et al (2005) Auxin inhibits endocytosis and promotes its own efflux from cells. Nature 435:1251. https://doi.org/10.1038/nature03633

Pan X, Chen J, Yang Z (2015) Auxin regulation of cell polarity in plants. Curr Opin Plant Biol 28:144–153

Paponov IA, Dindas J, Krol E, Friz T, Budnyk V, Teale W et al (2018) Auxin-induced plasma membrane depolarization is regulated by auxin transport and not by AUXIN BINDING PROTEIN1. Front Plant Sci 9:1953. https://doi.org/10.3389/fpls.2018.01953

Park E, Nebenfuhr A (2013) Myosin XIK of Arabidopsis thaliana accumulates at the root hair tip and is required for fast root hair growth. PLoS One 8(10):e76745. https://doi.org/10.1371/journal.pone.0076745

Parton R, Fischer-Parton S, Watahiki M, Trewavas AJ (2001) Dynamics of the apical vesicle accumulation and the rate of growth are related in individual pollen tubes. J Cell Sci 114 (14):2685–2695

Paul A, Pollard T (2008) The role of the FH1 domain and profilin in formin-mediated actin-filament elongation and nucleation. Curr Biol 18(1):9–19

Pei W, Du F, Zhang Y, He T, Ren H (2012) Control of the actin cytoskeleton in root hair development. Plant Sci 187:10–18

Pellegrin S, Mellor H (2005) The Rho family GTPase Rif induces filopodia through mDia2. Curr Biol 15(2):129–133

Peremyslov VV, Prokhnevsky AI, Avisar D, Dolja VV (2008) Two class XI myosins function in organelle trafficking and root hair development in Arabidopsis. Plant Physiol 146(3):1109–1116

Peremyslov VV, Prokhnevsky AI, Dolja VV (2010) Class XI myosins are required for development, cell expansion, and F-Actin organization in Arabidopsis. Plant Cell 22(6):1883–1897. https://doi.org/10.1105/tpc.110.076315

Peremyslov VV, Mockler TC, Filichkin SA, Fox SE, Jaiswal P, Makarova KS et al (2011) Expression, splicing, and evolution of the myosin gene family in plants. Plant Physiol 155 (3):1191–1204. https://doi.org/10.1104/pp.110.170720

Petroski MD, Deshaies RJ (2005) Function and regulation of cullin–RING ubiquitin ligases. Nat Rev Mol Cell Biol 6:9 (Supplementary information). https://doi.org/10.1038/nrm1547. https://www.nature.com/articles/nrm1547

Pierson E, Miller D, Callaham D, Van Aken J, Hackett G, Hepler P (1996) Tip-localized calcium entry fluctuates during pollen tube growth. Dev Biol 174(1):160–173

Pitts RJ, Cernac A, Estelle M (1998) Auxin and ethylene promote root hair elongation in Arabidopsis. Plant J 16(5):553–560

Pollard TD (2007) Regulation of actin filament assembly by Arp2/3 complex and formins. Annu Rev Biophys Biomol Struct 36:451–477

Pollard TD, Borisy GG (2003) Cellular motility driven by assembly and disassembly of actin filaments. Cell 112(4):453–465

Pollard TD, Mooseker MS (1981) Direct measurement of actin polymerization rate constants by electron microscopy of actin filaments nucleated by isolated microvillus cores. J Cell Biol 88 (3):654–659

Qin T, Liu X, Li J, Sun J, Song L, Mao T (2014) Arabidopsis microtubule-destabilizing protein 25 functions in pollen tube growth by severing actin filaments. Plant Cell 26(1):325–339. https://doi.org/10.1105/tpc.113.119768

Qu X, Zhang H, Xie Y, Wang J, Chen N, Huang S (2013) Arabidopsis villins promote actin turnover at pollen tube tips and facilitate the construction of actin collars. Plant Cell 25 (5):1803–1817. https://doi.org/10.1105/tpc.113.110940

Quintana-Cataño CA, Staiger CJ, Zhang W (2016) In vitro motility of actin filaments powered by plant myosins XI

Rahman A, Bannigan A, Sulaman W, Pechter P, Blancaflor EB, Baskin TI (2007) Auxin, actin and growth of the *Arabidopsis thaliana* primary root. Plant J 50(3):514–528

Ramachandran S, Christensen HE, Ishimaru Y, Dong C-H, Chao-Ming W, Cleary AL et al (2000) Profilin plays a role in cell elongation, cell shape maintenance, and flowering in Arabidopsis. Plant Physiol 124(4):1637–1647

Raudaskoski M, Åström H, Laitiainen E (2001) Pollen tube cytoskeleton: structure and function. J Plant Growth Regul 20(2):113–130

Reddy AS, Day IS (2001) Analysis of the myosins encoded in the recently completed *Arabidopsis thaliana* genome sequence. Genome Biol 2(7):research0024. 0021

Ren H, Xiang Y (2007) The function of actin-binding proteins in pollen tube growth. Protoplasma 230(3–4):171–182

Ridley AJ (2015) Rho GTPase signalling in cell migration. Curr Opin Cell Biol 36:103–112

Riedl J, Crevenna AH, Kessenbrock K, Yu JH, Neukirchen D, Bista M et al (2008) Lifeact: a versatile marker to visualize F-actin. Nat Methods 5(7):605

Ringli C, Baumberger N, Diet A, Frey B, Keller B (2002) ACTIN2 is essential for bulge site selection and tip growth during root hair development of Arabidopsis. Plant Physiol 129 (4):1464–1472

Robert S, Kleine-Vehn J, Barbez E, Sauer M, Paciorek T, Baster P et al (2010) ABP1 mediates auxin inhibition of clathrin-dependent endocytosis in Arabidopsis. Cell 143(1):111–121

Romagnoli S, Cai G, Cresti M (2003) In vitro assays demonstrate that pollen tube organelles use kinesin-related motor proteins to move along microtubules. Plant Cell 15(1):251–269

Romero S, Le Clainche C, Didry D, Egile C, Pantaloni D, Carlier M-F (2004) Formin is a processive motor that requires profilin to accelerate actin assembly and associated ATP hydrolysis. Cell 119(3):419–429

Rück A, Palme K, Venis MA, Napier RM, Felle HH (1993) Patch-clamp analysis establishes a role for an auxin binding protein in the auxin stimulation of plasma membrane current in *Zea mays* protoplasts. Plant J 4(1):41–46

Ruegger M, Dewey E, Gray WM, Hobbie L, Turner J, Estelle M et al (1998) The TIR1 protein of Arabidopsis functions in auxin response and is related to human SKP2 and yeast Grr1p. Genes Dev 12(2):198–207

Saarikangas J, Zhao H, Lappalainen P (2010) Regulation of the actin cytoskeleton-plasma membrane interplay by phosphoinositides. Physiol Rev 90(1):259–289. https://doi.org/10.1152/physrev.00036.2009

Sakata T, Oshino T, Miura S, Tomabechi M, Tsunaga Y, Higashitani N et al (2010) Auxins reverse plant male sterility caused by high temperatures. Proc Natl Acad Sci USA 107(19):8569–8574

Sandalio LM, Rodríguez-Serrano M, Romero-Puertas MC (2016) Leaf epinasty and auxin: a biochemical and molecular overview. Plant Sci 253:187–193

Sheahan MB, Staiger CJ, Rose RJ, McCurdy DW (2004) A green fluorescent protein fusion to actin-binding domain 2 of Arabidopsis fimbrin highlights new features of a dynamic actin cytoskeleton in live plant cells. Plant Physiol 136(4):3968–3978

Smith LG, Oppenheimer DG (2005) Spatial control of cell expansion by the plant cytoskeleton. Annu Rev Cell Dev Biol 21:271–295

Sparkes I (2011) Recent advances in understanding plant myosin function: life in the fast lane. Mol Plant 4(5):805–812

Staiger CJ, Poulter NS, Henty JL, Franklin-Tong VE, Blanchoin L (2010) Regulation of actin dynamics by actin-binding proteins in pollen. J Exp Bot 61(7):1969–1986

Steffens B, Feckler C, Palme K, Christian M, Böttger M, Lüthen H (2001) The auxin signal for protoplast swelling is perceived by extracellular ABP1. Plant J 27(6):591–599

Stephan OOH (2017) Actin fringes of polar cell growth. J Exp Bot 68(13):3303–3320. https://doi.org/10.1093/jxb/erx195

Stephan O, Cottier S, Fahlén S, Montes-Rodriguez A, Sun J, Eklund DM et al (2014) RISAP is a TGN-associated RAC5 effector regulating membrane traffic during polar cell growth in tobacco. Plant Cell 26(11):4426–4447. https://doi.org/10.1105/tpc.114.131078

Su H, Zhu J, Cai C, Pei W, Wang J, Dong H et al (2012a) FIMBRIN1 is involved in lily pollen tube growth by stabilizing the actin fringe. Plant Cell 24(11):4539–4554. https://doi.org/10.1105/tpc.112.099358

Su H, Zhu J, Cai C, Pei W, Wang J, Dong H et al (2012b) FIMBRIN1 is involved in lily pollen tube growth by stabilizing the actin fringe. Plant Cell 24(11):4539–4554. https://doi.org/10.1105/tpc.112.099358

Su H, Feng H, Chao X, Ding X, Nan Q, Wen C et al (2017) Fimbrins 4 and 5 act synergistically during polarized pollen tube growth to ensure fertility in Arabidopsis. Plant Cell Physiol 58 (11):2006–2016. https://doi.org/10.1093/pcp/pcx138

Suarez C, Carroll RT, Burke TA, Christensen JR, Bestul AJ, Sees JA et al (2015) Profilin regulates F-actin network homeostasis by favoring formin over Arp2/3 complex. Dev Cell 32(1):43–53

Sweeney HL, Houdusse A (2010) Structural and functional insights into the myosin motor mechanism. Annu Rev Biophys 39:539–557

Tan X, Calderon-Villalobos LIA, Sharon M, Zheng C, Robinson CV, Estelle M et al (2007) Mechanism of auxin perception by the TIR1 ubiquitin ligase. Nature 446:640. https://doi.org/10.1038/nature05731

Taylor LP, Hepler PK (1997) Pollen germination and tube growth. Annu Rev Plant Physiol Plant Mol Biol 48(1):461–491

Tominaga M, Kojima H, Yokota E, Nakamori R, Anson M, Shimmen T et al (2012) Calcium-induced mechanical change in the neck domain alters the activity of plant myosin XI. J Biol Chem 287(36):30711–30718

Traas J, Braat P, Emons A, Meekes H, Derksen J (1985) Microtubules in root hairs. J Cell Sci 76 (1):303–320

Ueda H, Yokota E, Kutsuna N, Shimada T, Tamura K, Shimmen T et al (2010) Myosin-dependent endoplasmic reticulum motility and F-actin organization in plant cells. Proc Natl Acad Sci USA 107(15):6894–6899. https://doi.org/10.1073/pnas.0911482107

Ueda H, Tamura K, Hara-Nishimura I (2015) Functions of plant-specific myosin XI: from intra-cellular motility to plant postures. Curr Opin Plant Biol 28:30–38

Van Gestel K, Slegers H, Von Witsch M, Samaj J, Baluska F, Verbelen JP (2003) Immunological evidence for the presence of plant homologues of the actin-related protein Arp3 in tobacco and maize: subcellular localization to actin-enriched pit fields and emerging root hairs. Protoplasma 222(1–2):45–52. https://doi.org/10.1007/s00709-003-0004-8

van Gisbergen PA, Li M, Wu S-Z, Bezanilla M (2012) Class II formin targeting to the cell cortex by binding PI (3, 5) P2 is essential for polarized growth. J Cell Biol 198(2):235–250

Vaškebová L, Šamaj J, Ovečka M (2017) Single-point ACT2 gene mutation in the Arabidopsis root hair mutant der1-3 affects overall actin organization, root growth and plant development. Ann Bot 122(5):889–901

Vidali L, Hepler PK (1997) Characterization and localization of profilin in pollen grains and tubes of Lilium longiflorum. Cell Motil Cytoskeleton 36(4):323–338. https://doi.org/10.1002/(sici)1097-0169(1997)36:4<323::Aid-cm3>3.0.Co;2-6

Vidali L, McKenna ST, Hepler PK (2001) Actin polymerization is essential for pollen tube growth. Mol Biol Cell 12(8):2534–2545

Vidali L, Rounds CM, Hepler PK, Bezanilla M (2009a) Lifeact-mEGFP reveals a dynamic apical F-actin network in tip growing plant cells. PLoS One 4(5):e5744

Vidali L, van Gisbergen PA, Guerin C, Franco P, Li M, Burkart GM et al (2009b) Rapid formin-mediated actin-filament elongation is essential for polarized plant cell growth. Proc Natl Acad Sci USA 106(32):13341–13346. https://doi.org/10.1073/pnas.0901170106

Villalobos LIAC, Lee S, De Oliveira C, Ivetac A, Brandt W, Armitage L et al (2012) A combinatorial TIR1/AFB–Aux/IAA co-receptor system for differential sensing of auxin. Nat Chem Biol 8(5):477

Voigt B, Timmers AC, Šamaj J, Müller J, Baluška F, Menzel D (2005) GFP-FABD2 fusion construct allows in vivo visualization of the dynamic actin cytoskeleton in all cells of Arabidopsis seedlings. Eur J Cell Biol 84(6):595–608

Wang B, Bailly A, Zwiewka M, Henrichs S, Azzarello E, Mancuso S et al (2013) Arabidopsis TWISTED DWARF1 functionally interacts with auxin exporter ABCB1 on the root plasma membrane. Plant Cell 25(1):202–214

Wu G, Otegui MS, Spalding EP (2010a) The ER-localized TWD1 immunophilin is necessary for localization of multidrug resistance-like proteins required for polar auxin transport in Arabidopsis roots. Plant Cell 22(10):3295–3304

Wu Y, Yan J, Zhang R, Qu X, Ren S, Chen N et al (2010b) Arabidopsis FIMBRIN5, an actin bundling factor, is required for pollen germination and pollen tube growth. Plant Cell 22 (11):3745–3763. https://doi.org/10.1105/tpc.110.080283

Wu S, Xie Y, Zhang J, Ren Y, Zhang X, Wang J et al (2015) VLN2 regulates plant architecture by affecting microfilament dynamics and polar auxin transport in rice. Plant Cell 27 (10):2829–2845. https://doi.org/10.1105/tpc.15.00581

Xiang Y, Huang X, Wang T, Zhang Y, Liu Q, Hussey PJ et al (2007) ACTIN BINDING PROTEIN29 from Lilium pollen plays an important role in dynamic actin remodeling. Plant Cell 19(6):1930–1946

Xu T, Wen M, Nagawa S, Fu Y, Chen J-G, Wu M-J et al (2010) Cell surface-and rho GTPase-based auxin signaling controls cellular interdigitation in Arabidopsis. Cell 143(1):99–110

Xu T, Dai N, Chen J, Nagawa S, Cao M, Li H et al (2014) Cell surface ABP1-TMK auxin-sensing complex activates ROP GTPase signaling. Science 343(6174):1025–1028

Yang Z (1998) Signaling tip growth in plants. Curr Opin Plant Biol 1(6):525–530

Yang W, Ren S, Zhang X, Gao M, Ye S, Qi Y et al (2011) BENT UPPERMOST INTERNODE1 encodes the class II formin FH5 crucial for actin organization and rice development. Plant Cell 23(2):661–680. https://doi.org/10.1105/tpc.110.081802

Ye J, Zheng Y, Yan A, Chen N, Wang Z, Huang S et al (2009) Arabidopsis formin3 directs the formation of actin cables and polarized growth in pollen tubes. Plant Cell 21(12):3868–3884

Yi K, Guo C, Chen D, Zhao B, Yang B, Ren H (2005) Cloning and functional characterization of a formin-like protein (AtFH8) from Arabidopsis. Plant Physiol 138(2):1071–1082

Zhang H, Qu X, Bao C, Khurana P, Wang Q, Xie Y et al (2010) Arabidopsis VILLIN5, an actin filament bundling and severing protein, is necessary for normal pollen tube growth. Plant Cell 22(8):2749–2767. https://doi.org/10.1105/tpc.110.076257

Zhang Y, Xiao Y, Du F, Cao L, Dong H, Ren H (2011a) Arabidopsis VILLIN4 is involved in root hair growth through regulating actin organization in a Ca2+-dependent manner. New Phytol 190 (3):667–682

Zhang Z, Zhang Y, Tan H, Wang Y, Li G, Liang W et al (2011b) RICE MORPHOLOGY DETERMINANT encodes the type II formin FH5 and regulates rice morphogenesis. Plant Cell 23(2):681–700. https://doi.org/10.1105/tpc.110.081349

Zhang M, Zhang R, Qu X, Huang S (2016) Arabidopsis FIM5 decorates apical actin filaments and regulates their organization in the pollen tube. J Exp Bot 67(11):3407–3417. https://doi.org/10. 1093/jxb/erw160

Zheng Y, Xie Y, Jiang Y, Qu X, Huang S (2013) Arabidopsis actin-depolymerizing factor7 severs actin filaments and regulates actin cable turnover to promote normal pollen tube growth. Plant Cell 25(9):3405–3423. https://doi.org/10.1105/tpc.113.117820

Zhou Z, Shi H, Chen B, Zhang R, Huang S, Fu Y (2015) Arabidopsis RIC1 severs actin filaments at the apex to regulate pollen tube growth. Plant Cell 27(4):1140–1161. https://doi.org/10.1105/ tpc.114.135400

Zhu J, Geisler MJ (2015) Keeping it all together: auxin–actin crosstalk in plant development. J Exp Bot 66(16):4983–4998

Zhu L, Zhang Y, Kang E, Xu Q, Wang M, Rui Y et al (2013) MAP18 regulates the direction of pollen tube growth in Arabidopsis by modulating F-actin organization. Plant Cell 25 (3):851–867

Zhu J, Zwiewka M, Sovero V, di Donato M, Ge P, Oehri J et al (2016) TWISTED DWARF1 mediates the action of auxin transport inhibitors on actin cytoskeleton dynamics. Plant Cell 28 (4):930–948. https://doi.org/10.1105/tpc.15.00726

Chapter 8
Interactions Between the Plant Endomembranes and the Cytoskeleton

Pengfei Cao and Federica Brandizzi

Abstract In eukaryotic cells, the endomembrane system comprises the endoplasmic reticulum (ER), the vacuole, and several other types of membrane-enclosed compartments that share membrane origins and communicate with each other. These endomembrane compartments are indispensable for the cell and together exert essential cellular functions, such as intracellular membrane transport and secretion. In plant cells, the endomembrane compartments interact extensively with the cytoskeleton system, mainly the actin cytoskeleton, in concert with their dynamic biogenesis and movement. Recent studies have characterized conserved mechanisms and a set of plant-specific proteins that are involved in the endomembrane–cytoskeleton interactions. In this chapter, we review mechanisms of the interactions between plant cytoskeleton and the major endomembrane compartments in a broad context of organelle morphogenesis, dynamics, and cellular functions.

8.1 Introduction

For most of the plant endomembrane compartments characterized to date, interactions with the cytoskeleton have been reported being necessary for their dynamic organization. Arguably, one of the best-characterized organelles for interactions with the cytoskeleton is the endoplasmic reticulum (ER), the first organelle of the endomembrane system, which is responsible for the production of secretory proteins and lipids. As the most extensive organelle that spans throughout the cytoplasm to connect other organelles from the nuclear envelope to the plasma membrane (Valm et al. 2017), the ER supports the dynamics and functions of most other organelles. In plant cells, the ER network forms thick membrane strands through tight interactions with F-actin bundles (Ueda et al. 2010), providing a highway for rapid intracellular transport that bypasses a jammed cytoplasm and even the large central vacuole

---

P. Cao · F. Brandizzi (✉)
MSU-DOE Plant Research Laboratory, Michigan State University, East Lansing, MI, USA

Department of Plant Biology, Michigan State University, East Lansing, MI, USA
e-mail: fb@msu.edu

© Springer Nature Switzerland AG 2019
V. P. Sahi, F. Baluška (eds.), *The Cytoskeleton*, Plant Cell Monographs 24,
https://doi.org/10.1007/978-3-030-33528-1_8

(Stefano et al. 2014a). In the first section of this chapter, we focus on the roles of cytoskeleton system in ER morphogenesis, rearrangement of ER tubules, and ER streaming, review recently identified plant proteins that are involved in ER–cytoskeleton interaction, and then provide ER–PM contact sites as an example of organelle–organelle interactions that rely on the cytoskeleton.

In most mature plant cells, the vacuole is the largest organelle that typically occupies more than 90% of the cell volume (Zhang et al. 2014). Pumping up the large-volume vacuoles provides turgor pressure to drive cell expansion and ultimately plant growth. Vacuoles are also the major storage compartment for proteins, sugars, along with many primary and secondary metabolites. Recent studies have revealed essential roles of the lytic vacuole in autophagy to degrade and recycle proteins, membrane lipids, and even organelles (Yang and Bassham 2015). Since these vacuolar activities are essential for plant development, mutations that arrest vacuole biogenesis lead to embryo lethality (Zhang et al. 2014). Evidence suggests that the vacuolar membranes are associated with the actin cytoskeleton, which is crucial for the morphology and biogenesis of plant vacuoles (Staiger et al. 1994; Kutsuna et al. 2003; Hoffmann and Nebenführ 2004; Higaki et al. 2006; Sheahan et al. 2007).

In addition to the ER and the vacuole, several other relatively smaller endomembrane compartments are connected with the plant cytoskeleton. In the plant cell secretory pathway, ER-produced cargoes are transported to the Golgi apparatus, which is dispersed into mini stacks, is closely associated with the ER, and moves along actin. The *trans*-Golgi network (TGN) is the sorting station for secretion/exocytosis, endocytosis, and vesicle transport to the vacuole. In plant cells, subpopulations of the TGN can be spatially independent from the Golgi apparatus. Their dynamics have been associated with both actin and microtubules (Sparks et al. 2016; Renna et al. 2018), and the underlying mechanisms will be discussed in this chapter. Additionally, autophagosomes are special transport vesicles that are formed in the autophagic response for degradation and recycling. The interplay between autophagosome and cytoskeleton system in mammalian and plant cells will be discussed.

In addition to reviewing various well-established and recently identified mechanisms of the endomembrane–cytoskeleton interactions in plant cells, we bring up topics of great interest for future research. First, the current understanding of the endomembrane–cytoskeleton interactions in plant cells is still limited in terms of both motor-driven dynamics and direct anchoring, which is also not well understood in yeast and mammalian cells. Furthermore, little is known about the signaling pathways beyond any of these direct interacting mechanisms that can integrate cytoskeleton remodeling and endomembrane dynamics to fulfill cellular and physiological functions. Brief summaries of potential regulatory mechanisms and plant-specific activities related to the endomembrane–cytoskeleton interactions are included for a constructive discussion of this topic.

8.2   ER

8.2.1   Cytoskeleton System Regulates Distribution of ER Tubules, Sheets, and Strands

A classical view of the ER network consists of sheet-like cisternae and tubules. Based mainly on models gathered in mammalian cells, formation of tubular ER is mainly dependent on transmembrane proteins with unique wedge-like topology, named reticulons and REEPs, which can bend the ER membrane and stabilize the curvature (Voeltz et al. 2006; Shibata et al. 2010). Tubulated ER undergoes homotypic fusion through the action of the dynamin-like GTPase atlastins (ATLs) and a Rab GTPase, Rab10 (Hu et al. 2009; English and Voeltz 2013). Additionally, Lunapark (Lnp) proteins stabilize the three-way junctions between ER tubules by interacting with reticulons, REEPs, and ATLs (Chen et al. 2012; Zhou et al. 2019). Studies in plant cells have functionally characterized *Arabidopsis* reticulons, Lnps, and the ATL homolog RHD3 (Tolley et al. 2008; Sparkes et al. 2010; Chen et al. 2011; Stefano et al. 2012; Lee et al. 2013; Zhang et al. 2013; Kriechbaumer et al. 2015, 2018; Breeze et al. 2016; Ueda et al. 2018). Four mechanisms may contribute to shaping ER sheets: (1) ER membrane-bound ribosomes may flatten and stabilize ER sheets; (2) transmembrane proteins with a large coiled-coil domain in ER lumen, such as CLIMP63 in mammalian cells, can self-interact and bring the opposite membranes together, serving as luminal spacers; (3) transmembrane proteins with a large coiled-coil domain in cytosol, for example, p180 and kinectin in mammalian cells, can self-interact on the cytosolic side and stabilize the flat ER sheet; and (4) membrane-bending reticulon and REEP proteins curve the edges of ER sheet (Shibata et al. 2009; Westrate et al. 2015).

Multiple lines of evidence suggest that the establishment of ER tubules and sheets is essentially determined by these membrane-shaping proteins. First, a tubular membrane network can be reconstituted in vitro simply with lipids and a small set of conserved membrane-shaping proteins (Wang et al. 2016c; Powers et al. 2017). Second, overexpression or ablation of membrane-shaping proteins affects the ER sheet-tubule balance in plant cells (Tolley et al. 2008; Sparkes et al. 2010; Chen et al. 2011; Stefano et al. 2012; Zhang et al. 2013; Kriechbaumer et al. 2018; Ueda et al. 2018). Third, mutant screens for regulators of ER morphology identified proteins localized to ER membrane and secretory pathway, but not the cytoskeleton system (Prinz et al. 2000; Chen et al. 2012; Stefano et al. 2012). Additionally, mathematical modeling of ER morphogenesis indicates that the influence of cytoskeletal forces seems to be limited to the spatial distribution of ER membranes (Shemesh et al. 2014). Collectively, this evidence prompted the hypothesis that the abundance of different membrane-shaping proteins and their interactions with membrane lipids form an ER network with sheets and tubules (Shemesh et al. 2014). Nonetheless, the established ER tubules and sheets are distributed around the cell by cytoskeletal forces. The ER network in animal cells is mainly regulated by microtubules. When chemical intervention triggers depolymerization of microtubules, peripheral ER tubules retract to the central nuclear envelope and form collapsed ER sheets (Terasaki et al. 1986; Shibata et al. 2008). In yeast cells, the integrity of actin cytoskeleton, rather than microtubules, is required for ER movement and its morphology (Prinz et al. 2000; Fehrenbacher et al. 2002). For plant cells, early studies using fluorescence and electron microscopies reported that ER membranes are frequently distributed in close proximity to actin cables (Quader et al. 1989; Quader 1990; Staehelin 1997). Disruption of microtubules does not significantly alter the ER morphology, while both treatment of actin-depolymerizing drugs and genetic ablation of myosin motors lead to collapsed ER with enlarged ER sheets and disappeared ER strands (Ueda et al. 2010). ER strands are a prominent feature of the plant ER, which is established by interactions between ER membranes and thick actin bundles (Staehelin 1997; Boevink et al. 1998; Peremyslov et al. 2010; Ueda et al. 2010; Stefano et al. 2014b; Cao et al. 2016). Disrupting thick actin bundles by latrunculin B treatment or knocking out class XI myosin genes leads to disappearance of ER strands and enlarged ER sheets (Peremyslov et al. 2010; Ueda et al. 2010). Another example is mutation of SYP73, an anchor of ER membrane to actin filament, which phenocopies disruption of the actin–myosin system (Cao et al. 2016). By contrast, plant cells stimulated to form extreme actin bundling display thickened and elongated ER strands (Cao et al. unpublished data). Taken together, these studies suggest that the ER strands are an outcome of enhanced actin bundling and ER membrane–actin interactions.

The nature of ER sheets and ER strands in plant cells is still unclear. Recent studies of ER in cultured mammalian cells developed spatiotemporal analyses with super-resolution microscopy, which revealed that most peripheral ER structures classically identified as ER sheets are actually dense and dynamic matrices of ER tubules (Nixon-Abell et al. 2016; Guo et al. 2018). Compared to cultured mammalian cells, commonly adopted plant cell types for ER study, such as tobacco leaf epidermal cells, *Arabidopsis* hypocotyl epidermal cells, and expanding cotyledon epidermal cells, generally display ER networks with a larger portion of sheets (Sparkes et al. 2009a; Ueda et al. 2010; Stefano et al. 2014b). Meanwhile, these plant cell types are highly vacuolated so that the cytoplasmic content is pressured into a thin layer. Therefore, it is plausible, though lacking direct evidence so far, that in plant cells the ER sheets and ER strands are also dense matrices of tubules, and on the other hand, the formation of isolated ER tubules and thick ER strands in plant cells is largely dependent on the availability of cytoplasmic space sustained by cytoskeleton and other organelles.

8.2.2   Rearrangement of the Tubular ER

In mature cells, such as expanded *Arabidopsis* cotyledon epidermal cells, the ER network generally does not experience dramatic alteration of morphology (Stefano et al. 2014b); nevertheless, the entire ER network consistently undergoes fine rearrangement of ER tubules. Interestingly, despite the ER networks in mammalian cells and plant cells being primarily bound to two distinct types of cytoskeleton cables and motors, the ER tubules move with seemingly similar mechanisms. Studies in mammalian cells identified five scenarios of tubular ER initiation and elongation. In the three most common situations, the tubular ER is driven by a strong association with the cytoskeleton system. The tip of an ER tubule can be pulled by kinesin or dynein motor sliding on a static microtubule, or alternatively form a Tip Attachment Complex (TAC) with the plus end of a microtubule that is being polymerized or depolymerized (Waterman-Storer and Salmon 1998; Bola and Allan 2009; Westrate et al. 2015; Guo et al. 2018). In other cases, ER tubules can be pulled out by moving organelles, such as endosomes and lysosomes, or initiate de novo ER tubulation without any nearby microtubules or small organelles (Guo et al. 2018).

In plant cells, the actin cytoskeleton controls the organization and dynamics of ER network, including tubular ER movement. Studies showed that latrunculin B-induced depolymerization of actin filaments halts tubular ER generation (Sparkes et al. 2009a; Yokota et al. 2011). Furthermore, simultaneous imaging of ER tubules and actin bundles in plant cell extracts identified tubular ER elongation along the actin bundles, presumably driven by myosin (Yokota et al. 2011). However, compared to microtubules, actin filaments are thinner, more dynamic, and forming a densely packed network, which is challenging for high-resolution imaging even in thin-layer cultured mammalian cells (Guo et al. 2018; Pollard and Goldman 2018). It remains elusive whether in plant cells the ER tubule tip can bind to and be pulled by an end of polymerizing or depolymerizing actin filament, which would be resembling the tip binding between tubular ER and microtubule in mammalian cells. Advances of imaging techniques may also elucidate the mechanisms of microfilament-dependent ER reorganization in animal cells (Poteryaev et al. 2005; Joensuu et al. 2014) and unveil whether single actin filaments are involved in the de novo ER tubulation.

The microtubule cytoskeleton plays a minor role in ER dynamics. Recent studies closely monitored ER rearrangement in *Arabidopsis* hypocotyl epidermal cells and observed motor-driven ER tubule extension along microtubules, as well as ER subdomains that are statically anchored to crossing microtubules (Hamada et al. 2014). This work did not record ER tubule extension associated with microtubule growing plus end, suggesting that the tip-attached TAC mechanism is not conserved in plant cells (Hamada et al. 2014). The mammalian TAC includes the microtubule plus end-binding protein EB1 and the ER transmembrane protein STIM1 (Westrate et al. 2015). AtEB1 exerts conserved functions of binding to microtubule plus end and regulating microtubule organization (Mathur et al. 2003; Dhonukshe et al. 2005; Molines et al. 2018). On the one hand, AtEB1 exhibits dual localization to microtubules and elusive endomembranes (Mathur et al. 2003), and it is involved in modulation of endocytosis (Dhonukshe et al. 2005). On the other hand, the functional homolog of STIM1 in plants has not been identified yet. Taken together, these studies suggest that plant tubular ER rearrangement along microtubules mainly rely on microtubule motor proteins.

8.2.3 ER Streaming

Imaging analyses of ER-retained fluorescent proteins report ER streaming as rapid bulk flow inside the ER lumen (Boevink et al. 1996a; Ueda et al. 2010; Stefano et al. 2012, 2014b). Previous studies suggest that ER streaming is largely dependent on the integrity of ER strands. When the thick ER strands are abolished and the ER collapses due to depletion of myosin XI proteins or the ER membrane-actin anchoring protein SYP73, the maximal and average velocities of ER streaming are reduced (Ueda et al. 2010; Cao et al. 2016). On the other hand, strikingly enhanced ER strands caused by a mutation of RHD3, the plant ATL homolog shaping ER tubules, also leads to significantly reduced ER streaming (Stefano et al. 2012). Therefore, both functional myosin XI motors and ER–cytoskeleton anchors are required for rapid ER streaming.

In plant cells, myosin XI proteins organize actin filaments and transport organelles on the actin tracks (Ueda et al. 2015; Nebenfuhr and Dixit 2018). A characterization of myosin XI members using live cell imaging and single and high-order mutants suggests that certain myosin XIs display ambiguous localizations and are involved in movement of multiple organelles (Avisar et al. 2008; Peremyslov et al. 2008; Prokhnevsky et al. 2008; Ueda et al. 2010; Griffing et al. 2014; Madison et al. 2015). Myosin XI-K has been suggested to be associated with ER membrane, and it exerts a dominant role for moving ER network (Ueda et al. 2010; Griffing et al. 2014). However, it is still under debate whether the ER is the cargo and the cellular site of binding for myosin XI-K. Two lines of evidence are in conflict with this. First, manipulation of myosin proteins strongly affects the dynamics of other small organelles such as Golgi, peroxisomes, mitochondria, and endosomes (Avisar et al. 2008; Prokhnevsky et al. 2008). Besides, more than a dozen of plant myosin XI-binding proteins that have been identified so far are distributed to disperse punctae in the cytoplasm, rather than resembling a typical ER network (Peremyslov et al. 2012, 2013; Kurth et al. 2017). Given that the ER is a complex organelle with distinct subdomains (Staehelin 1997; Stefano and Brandizzi 2018), it is possible that the myosin XI-binding proteins are distributed only over specific ER subdomains. Considering the extensive and strong association of the ER network with smaller organelles and endosomes in plant cells (Stefano et al. 2015), it is also possible that myosin XIs drive small organelles and indirectly propel ER movement. Identification of new myosin adaptors and advance of imaging techniques may further our understanding of the relationship between ER movement and myosin motors.

8.2.4 Recently Characterized Plant Proteins Involved in ER-Actin Interactions

The NETWORKED (NET) superfamily proteins contain an N-terminal NET-actin-binding (NAB) domain and a coiled-coil domain that is responsible for protein–protein interactions and protein targeting to specific membrane structures (Deeks et al. 2012). Accordingly, three NET3 family members expressed in plant cells show filamentous distribution in colocalization with actin filaments and brighter foci that colocalize with distinct membranes. NET3A is accumulated around the nuclear membrane, whereas NET3B and NET3C are associated with the ER network. NET3C interacts with membrane-associated VAP27 in close proximity to the plasma membrane. Interestingly, VAP27 also interacts with microtubules (Wang et al. 2014, 2016a). These results imply that NET3C and VAP27 define ER–PM contact sites that incorporate both F-actin and microtubule cytoskeletons. Compared with NET3C, NET3B appears to be more tightly associated with the ER since overexpression of NET3B modulates the ER morphology in a manner reminiscent of SYP73-rearranged ER over the actin cables. Therefore, NET3B may function in anchoring ER network to the cytoplasmic F-actin. This feature is supported by the observation that overexpression of NET3B restricts ER membrane diffusion (Wang and Hussey 2017). Further analyses are expected to reveal how the peripheral membrane protein NET3B is connected to the ER membrane and whether the NET3-proteins interact directly with actin.

SYP73 was identified in search of ER-localized proteins with both transmembrane domain and coiled-coil domain, which may anchor ER membrane to cytoskeleton as CLIMP63, p180, and kinetin function in mammalian cells (Cao et al. 2016). However, the coiled-coil domain of SYP73 is cytosolic and adjacent to an unconventional actin-binding domain, hence forming strong self-interactions that presumably strengthens actin binding rather than spacing or flattening the ER membrane (Cao et al. 2016). As a consequence, overexpression of SYP73 leads to a rearrangement of ER network over actin cables (Cao et al. 2016), rather than augmenting ER sheets as it occurs for overexpression of CLIMP63 or p180 in mammalian cells (Shibata et al. 2010).

Our current understanding of ER-associated actin-binding proteins may be still very limited. A study of cotton fibers identified GhCFE1A as a protein linking the ER membrane and actin cytoskeleton that functions in cotton fiber initiation and elongation (Lv et al. 2015). Surprisingly, a recent study in mammalian cells revealed that Sec61β, a very commonly used marker of ER membrane in mammalian cells, binds microtubules and is able to remodel the network to overlay microtubules upon overexpression (Zhu et al. 2018b). More proteins involved in ER–cytoskeleton interactions are expected to be identified with advances of organelle proteomics and methods studying protein–protein interactions (Kriechbaumer et al. 2015; Wang et al. 2017b; Hoyer et al. 2018).

8.2.5   Membrane–Cytoskeleton Interaction at the ER–PM Contact Site

The ER network spreads throughout the cytoplasm and interacts most other membrane-bound organelles to form inter-organelle contact sites. One type of contact sites that has gained recent attention in plant research is between the ER and the plasma membrane (PM). The ER–PM contact sites (EPCS) have been implicated in endomembrane dynamics, cytoskeleton organization, and intercellular communication as well as in functioning as a signaling platform to respond to biotic and abiotic stresses (Bayer et al. 2017; Wang et al. 2017a; Stefano et al. 2018).

Several conserved protein families that bring ER and PM membranes together have been characterized in *Arabidopsis*, including extended-synaptotagmin family proteins (E-Syt, known as Syt1 in plants) and VAMP-associated proteins (VAP) (Saravanan et al. 2009; Lewis and Lazarowitz 2010; Wang et al. 2014, 2016a; Siao et al. 2016; Stefano et al. 2018; Lee et al. 2019). VAP27 directly binds to microtubules, and it also interacts with NET3C, a member of the plant-specific membrane-associated actin-binding NET family (Wang et al. 2014).

In mammalian cells, both microtubules and actin cytoskeleton regulate the EPCS. A store-operated calcium entry (SOCE) complex is formed by plasma membrane-localized protein Orai1 and ER transmembrane protein STIM1, which can interact with the microtubule plus end-binding protein EB1 to regulate the calcium gating function of SOCE at the EPCS (Chang et al. 2018). Recent studies reported interplay between actin organization and the dynamics of EPCS (del Dedo et al. 2017; van Vliet et al. 2017). A major type of the EPCS is supported by ER-associated VAP protein interacting with ORP, which also binds to phospholipids on the PM (Henne et al. 2015). A recent study in yeast identified an additional interaction between ORPs and myosin-I proteins (del Dedo et al. 2017). Upon disruption of the EPCS, the local actin polymerization is impaired and the formation of endocytic vesicles is stalled (del Dedo et al. 2017; Stefano et al. 2018). Interestingly, the actin cytoskeleton also transduces ER luminal signal to regulate the EPCS. Secretory proteins are produced in ER, and disturbance of proteostasis in the ER lumen triggers the unfolded protein response (UPR) that is detected by ER transmembrane sensors. A mammalian UPR sensor protein PERK interacts with an actin-binding protein, filamin A, to transduce the UPR signal to actin reorganization underneath the plasma membrane domain of EPCS (van Vliet et al. 2017). Consistently, functional PERK is required for actin dynamics and EPCS formation (van Vliet et al. 2017). Although plant homologs of PERK and filamin A have not been reported yet, the functional conservation of ER stress response mechanisms and F-actin side-binding proteins throughout eukaryotes suggests the existence of such interplay between the actin cytoskeleton and membranes at EPCS (Meagher and Fechheimer 2003; Angelos et al. 2017).

8.3 Vacuole

8.3.1 The Vacuolar Membrane Is Associated with the Cytoskeleton

Earlier studies of the plant vacuole and the cytoskeleton suggested that these structures are associated, or at least, distributed in close proximity. Isolated and cultured cells, such as tobacco BY-2 cells and *Arabidopsis* protoplasts, were first utilized to visualize the vacuole membranes and the actin cytoskeleton (Staiger et al. 1994; Kutsuna et al. 2003; Hoffmann and Nebenführ 2004; Higaki et al. 2006; Sheahan et al. 2007). Disruption of the actin cytoskeleton alters the vacuole morphology and the structures formed by the vacuolar membranes, most notably the transvacuolar strands (TVS). Upon microinjection of profilin, a potent actin monomer-binding protein that leads to F-actin depolymerization, TVS were deformed and cytoplasmic streaming stagnated (Staiger et al. 1994). This pioneering research, based mainly on bright-field analyses, suggested that the actin cytoskeleton is involved in vacuole architecture, especially the TVS. The vacuolar structures can be easily visualized by GFP-tagged vacuole membrane-associated protein or FM4-64 and BCECF staining. Improved imaging in cultured cells revealed that the vacuole membranes form not only spherical structures but also tubules, vesicles, and other intricate configurations; moreover, the TVS appear to be composed of these minuscule yet complicated structures clustering along thick actin bundles (Kutsuna et al. 2003; Hoffmann and Nebenführ 2004; Sheahan et al. 2007; Szymanski and Cosgrove 2009). The TVS gradually disappear after application of actin polymerization inhibitors, while they are insensitive to microtubule-targeting chemicals (Higaki et al. 2006; Sheahan et al. 2007). Furthermore, certain myosin inhibitors induce TVS collapse, while they have moderate or insignificant effects on the integrity of actin filaments (Hoffmann and Nebenführ 2004; Sheahan et al. 2007). These lines of evidence suggest that the maintenance of TVS is dependent on actin filaments rather than microtubules, possibly via specific myosin motors and receptors. However, our current knowledge of TVS has been mostly acquired through genetic or chemical disruption of the cytoskeleton. How the TVS structures are established, what is their function, and whether they are dynamically regulated are still open questions. It is possible that the strands are essentially cytoplasmic actin bundles wrapped by vacuole membranes that are maintained by strong membrane–cytoskeleton anchoring. In this scenario, the cytoskeleton-associated vacuolar membranes are possibly remains of smaller vacuoles after they undergo homotypic membrane fusion to form the large central vacuole. It is also possible that the TVS are induced-to-form structures that are regulated by cytoskeleton signaling.

8.3.2 Plant Cytoskeleton Controls Vacuole Morphogenesis

Whereas young meristematic cells generated either in embryonic or growing tissues contain multiple types of vacuoles, most mature plant cells contain one large central vacuole (Feeney et al. 2018; Cui et al. 2019). Several studies in plant cells proposed the dynamic vacuole membrane fusion as the essence of vacuole biogenesis that is dependent on a remodeling of the actin cytoskeleton (Li et al. 2013; Zheng et al. 2014). In yeast cells, multiple vacuoles frequently undergo fusion and division in response to growth conditions and to exert their stress-related functions (Li and Kane 2009). Vacuole membrane fusion is achieved by three stages: (1) priming of membrane lipid composition, (2) tethering regulated by Rab GTPase, and then

(3) docking and fusion mediated by HOPS and SNAREs complexes (Wickner 2010). Particularly, the homeostasis of phosphoinositides, an essential type of regulatory lipids for membrane fusion, also signals to the WAVE complex to regulate actin reorganization (Eitzen 2003; Takenawa and Suetsugu 2007). In addition, yeast vacuole fusion requires Cdc42p and Rho1p, two Rho GTPases that classically regulate actin remodeling through WAVE-Arp2/3 complexes (Eitzen et al. 2001, 2002; Muller et al. 2001). During vacuole membrane fusion, Cdc42p is activated and, in turn, stimulates actin polymerization to facilitate membrane docking before fusion (Isgandarova et al. 2007; Jones et al. 2010; Bodman et al. 2015). Considering that both yeast and plant cells utilize actin rather than microtubules to mobilize organelles, and Rho of Plants (ROP) GTPases play conserved role of signaling to cytoskeleton remodeling (Feiguelman et al. 2018), it is reasonable to hypothesize that the mechanism of actin-dependent vacuole fusion is conserved in plant cells. This hypothesis is supported by studies in plant cells (Li et al. 2013; Zheng et al. 2014) although the underlying mechanisms remain elusive.

Vacuolar occupancy of the cellular space during plant cell growth is determined by the number and size of vacuoles in a cell. Despite that providing turgor pressure to stimulate cell expansion has been recognized as a classic and critical function of the plant vacuole (Taiz 1984), the mechanisms regulating the vacuolar occupancy of cellular space and thereby controlling cell expansion are still poorly understood. Recently, two studies explored the interplay between vacuolar occupancy, the actin cytoskeleton, and the major plant growth hormone auxin (Löfke et al. 2015; Scheuring et al. 2016). Vacuole dynamics, organization of actin cytoskeleton, and cell expansion were found to be individually affected by exogenously applied auxin in an auxin receptor-dependent manner (Löfke et al. 2015; Scheuring et al. 2016). Furthermore, chemical interference and mutations of the action–myosin system were found to mitigate the vacuolar occupancy of the cellular space induced by auxin (Scheuring et al. 2016). The downstream of auxin signaling is complex and multiple mechanisms may contribute to cell expansion; nevertheless, these studies provided significant evidence for vacuole–cytoskeleton interactions in response to developmental signals.

8.3.3 Actin-Binding Proteins Associated with Vacuole

Recent studies in searching actin-binding proteins yielded novel plant proteins that are involved in interactions between the vacuole and the actin cytoskeleton. As discussed above, the plant-specific NET family consists of actin-binding proteins that are associated with various membranes. In *Arabidopsis* root epidermal cells, native promoter-driven expression of NET4A-GFP was found distributed on filamentous structures that are closely attached to the vacuolar membrane (Deeks et al. 2012). A study of vacuolar-type H^+-ATPase (V-ATPase) B subunits identified a conserved actin-binding domain and confirmed their actin-binding ability and the effects on actin filament remodeling in vitro (Ma et al. 2012). Despite the vacuolar-type nomenclature, V-ATPase subunits are distributed throughout the secretory pathway and may not necessarily be contributing to the interaction between vacuolar membrane and actin cytoskeleton (Ma et al. 2012).

8.4 Other Endomembrane Compartments

8.4.1 Golgi

In mammalian cells, the Golgi stacks are linked in tandem to form a unique Golgi "ribbon," which is mostly distributed near the nucleus and the microtubule organizing center (MTOC). The positioning of the mammalian Golgi, as well as formation of Golgi "ribbon," is dependent on sophisticated interactions between Golgi matrix proteins and peripheral microtubules, microtubule-associated proteins, actin-binding proteins, and motor proteins (Gosavi and Gleeson 2017).

In plants, the Golgi apparatus is composed of numerous mobile Golgi mini-stacks composed of several flattened membrane cisternae. The Golgi stacks are attached to the ER network and receive secretory cargo directly from the ER (Boevink et al. 1998; Brandizzi et al. 2002; Sparkes et al. 2009b), independently of actin and microtubules (Brandizzi et al. 2002). The plant Golgi stacks also undergo long-range movement in a stop-and-go manner on the actin cables (Boevink et al. 1998; Nebenfuhr et al. 1999). Chemical disruption of F-actin, chemical inhibition of myosins, or genetic ablation of Myosin XI genes causes reduced mobility of the Golgi (Boevink et al. 1998; Nebenfuhr et al. 1999; Brandizzi et al. 2002; Peremyslov et al. 2008; Prokhnevsky et al. 2008; Avisar et al. 2012). By contrast, treatment of microtubule-disrupting chemicals does not affect Golgi movement (Nebenfuhr et al. 1999; Brandizzi et al. 2002).

Microtubules have not been considered required for long-range movement of the plant Golgi; however, examination of the diversified motor protein family in plants suggests that microtubule components are still involved in regulating Golgi dynamics. *Arabidopsis* and cotton homologs of Kinesin-13A localize to the Golgi (Lu et al. 2005; Wei et al. 2009). Since knocking out Kinesin-13A caused growth phenotypes and subtle aggregation of Golgi, it has been speculated that the kinesins may serve as temporal anchors between the Golgi and microtubules and regulate the Golgi positioning (Lu et al. 2005; Wei et al. 2009).

Microscopy and quantitative analyses have provided evidence supporting a regulatory interplay between local actin organization and the stochastic Golgi movement. It has been reported that in epidermal cells of the root elongation zone, the Golgi stacks undergo fast movement with a velocity up to 7 μm s^{-1} in areas where thick and long actin bundles appear, while the Golgi stacks move more slowly and wiggle where F-actin is less bundled (Akkerman et al. 2011). Another study using hypocotyl cells led to similar results showing a correlation between locally enhanced actin bundling and Golgi movement with higher velocity and less wiggly behavior (Breuer et al. 2017). Moreover, through the analyses of Golgi movement and actin organization, this study showed the potential of accurately predicting the movement pattern of overall Golgi stacks as well as the direction and velocity of specific Golgi movement (Breuer et al. 2017). The microscopy analyses facilitated by quantitative modeling (Akkerman et al. 2011; Breuer et al. 2017) suggest that the intracellular organelle movement can be orchestrated by signaling to actin reorganization, and therefore is neither static nor completely stochastic.

8.4.2   TGN

Succeeding the Golgi along the biosynthetic route, the *trans*-Golgi network (TGN) is an endomembrane compartment where the secretory pathway and the endocytic pathway converge. Owing to the presence of a cellulosic cell wall, the TGN in plant cells likely exerts non-conserved cargo sorting functions compared to animal cells. The plant TGN also exhibits unique dynamics and spatial distribution. In animal cells, the TGN is attached to the Golgi and collectively is transported along microtubules by microtubule-associated motors and anchoring proteins (Yadav and Linstedt 2011). The plant TGN population is composed of a subgroup that associates with the Golgi (GA-TGN/early TGN) and a subgroup that is isolated from the Golgi (GI-TGN/late TGN) (Uemura et al. 2014; Renna et al. 2018). In plant cells, chemical manipulation of the actin cytoskeleton and microscopic analyses suggested that the TGN is associated with and dependent on F-actin for transport (Baluska et al. 2002; Kim et al. 2005).

In this section, we will review recently identified proteins that directly connect the TGN to the cytoskeleton cables. For example, RISAP is a TGN-localized effector of RAC5 GTPase that can directly bind to F-actin and myosin XI, regulating membrane traffic and tip growth of tobacco pollen tubes (Stephan et al. 2014). RISAP contains a DUF593 domain that is shared by identified plant myosin adaptor proteins (Peremyslov et al. 2013; Stephan et al. 2014). Another TGN protein, HLB1, was recently identified in a forward screen for mutants whose primary root elongation is hypersensitive to latrunculin B (Sparks et al. 2016). Interacting with TGN-localized MIN7/BEN1 protein, HLB1 colocalizes with TGN markers and moves along F-actin cable, suggesting it functions at the intersection of post-Golgi trafficking and actin cytoskeleton (Sparks et al. 2016). The *hlb1* mutant exhibits broad but mostly mild defects in vesicular transport, including both exocytosis and endocytosis (Sparks et al. 2016). Meanwhile, in the *hlb1* mutant, the organization of F-actin is disrupted and displays less bundling in root epidermal cells and root hairs (Sparks et al. 2016). Therefore, HLB1 likely functionally connects the TGN to the actin cytoskeleton in a manner dependent on BEN1, although a direct interaction with HLB1 and actin has not been demonstrated yet. If confirmed, the results would lend support to the earlier findings that the TGN dynamics depend on actin (Baluska et al. 2002; Kim et al. 2005). A recent research identified a novel regulator of TGN dynamics, TGNap1, which is required for post-Golgi traffic to the cell surface and vacuole, microtubule-dependent biogenesis of a subpopulation of TGNs, as well as efficient endocytosis

(Renna et al. 2018). Mechanistically, TGNap1 interacts with Rab6 and YIP4, two regulators of vesicular trafficking that bind on and off the TGN; TGNap1 also directly binds microtubules (Renna et al. 2018). These results support the novel model whereby post-Golgi organelle traffic and biogenesis depend partially on microtubules. In this model, TGNap1 dynamically bridges the TGN to microtubules for biogenesis and traffic of a subpopulation of GI-TGNs (Renna et al. 2018). Further characterization of TGN-associated proteins and their potential interactions with cytoskeleton components are expected to provide a comprehensive understanding of TGN dynamics and its crucial functions as the cargo sorting station.

8.4.3   Endosomes and Endocytosis

Endosomes are small endomembrane compartments that internalize plasma membrane-associated proteins and carry cargo for recycling, degradation, or sequestration. This process includes endosome biogenesis, transport, and sorting, which are facilitated by the cytoskeleton. In animal cells, the biogenesis and intracellular movement of endosomes are mainly dependent on actin cytoskeleton and microtubules, respectively (Granger et al. 2014; Simonetti and Cullen 2019). On the plasma membrane, activation of WASH and Arp2/3 complexes initiates actin nucleation and polymerization to form a meshwork of actin filaments, which restricts a subdomain for endosome biogenesis (Simonetti and Cullen 2019). Subsequently, the long-range intracellular traffic of endosomes is assumed by microtubules and associated motors. More than a dozen of kinesin and dynein motor proteins have specific client cargoes and largely identified adaptor proteins (Granger et al. 2014). The cytoplasmic actin components are also frequently associated with endosomes and target membranes to control multiple aspects of the endosomes beyond biogenesis, including sorting and short-range transport of cargoes and subtle dynamics (Granger et al. 2014). Interestingly, actin polymerization on the endosome surface initiates and stabilizes tubulation of membranes and eventually leads to fission and myosin-driven transport. By contrast, studies of various endosomal cargoes in plant cells indicated that actin cytoskeleton is the primary driving force, though microtubules and microtubules-associated motors are also involved in endosomal transport (Beck et al. 2012; Ambrose et al. 2013; Kang et al. 2014; Rakusová et al. 2019).

A type of extensively investigated endosomal cargoes in plant cells are AUX1 and PIN auxin transporters, whose proper distribution on the plasma membrane of one side of the cell for directional auxin influx and efflux is crucial for the establishment of cell polarity as well as overall plant growth and development (Friml 2010). The transport and distribution of PIN1, PIN3, and AUX1 are all primarily dependent on the integrity of F-actin, rather than microtubules (Geldner et al. 2001; Friml et al. 2002; Kleine-Vehn et al. 2006; Hu et al. 2015). Other evidence supports that microtubules contribute to the traffic of PIN-associated endosomes (Ambrose et al. 2013). The microtubule-associated protein CLASP interacts with sorting nexin SNX1 and connects the endosomal membrane with microtubules (Ambrose et al. 2013). Knockout mutants of *CLASP* show disrupted polar distribution of PIN2 as well as auxin-related plant growth phenotypes (Ambrose et al. 2013). Furthermore, functional CLASP and integrity of microtubules prevent PIN2 from degradation (Ambrose et al. 2013). Nevertheless, it should be noted that in animal cells the SNX proteins are a type of key regulators of endosome sorting and traffic, and several associations between specific SNX and kinesin/dynein proteins regulate the traffic of distinct endosomal cargoes (Granger et al. 2014). Additional research also demonstrated the role of *Arabidopsis* SNXs in modulating the trafficking of the iron transporter IRT1 and vacuolar storage proteins (Niemes et al. 2010; Pourcher et al. 2010; Ivanov et al. 2014), suggesting conserved functions of *Arabidopsis* SNXs for sorting and trafficking. Interestingly, a recent forward genetic screen for novel regulators of PIN3 cellular distribution identified a new mutant allele of ACTIN2 and reported that disruption of F-actin and microtubules confers distinct effects on PIN3 polarity and gravitropism (Rakusová et al. 2019). Therefore, actin cytoskeleton and microtubules may be required for endosomal transport and assume diversified tasks.

Endocytosis facilitated by the actin cytoskeleton is critical for pathogen pattern recognition receptors (PRR) during plant immunity. Monitoring the mobilization of GFP-tagged FLAGELLIN SENSING2 (FLS2), a plasma membrane-localized receptor of bacterial flagellin (flg22), upon flg22 induction and chemical treatment revealed critical but differential roles of both F-actin and myosins (Beck et al. 2012). Treatment with the myosin inhibitor 2,3-butanedione monoxime largely reduced the formation of FLS2-GFP-labeled endosomes by approximately 80% compared to the control (Beck et al. 2012). By contrast, depolymerizing F-actin by latrunculin B treatment did not impede internalization of FLS2-GFP, but rather halted the FLS2-GFP-labeled endosomes from leaving the plasma membrane (Beck et al. 2012). Consistently, HopW1, a bacterial effector protein injected into the plant cell, disrupts F-actin and inhibits endocytosis (Kang et al. 2014). These results are critical pieces of evidence supporting a broad model in which dynamic interactions between the cytoskeleton and membrane-bound organelles perceive pathogen invasion signals and coordinate downstream subcellular responses (Li and Day 2019).

8.4.4 Other Transport Vesicles

The cellulose synthase complex (CSC) is an essential machinery for cell wall synthesis and biomass accumulation. It is well established that, at the plasma membrane, the CSC moves along the underlying cortical microtubule framework to synthesize and deposit cellulose microfibrils into the cell wall (Bashline et al. 2014). However, much less is known about the mechanisms of transporting the CSC from Golgi to plasma membrane. The plasma membrane loci, where the CSCs are delivered, are determined by the underlying cortical microtubules, though disruption of the actin cytoskeleton also affects distribution pattern of CSCs on the plasma membrane (Crowell et al. 2009; Gutierrez et al. 2009). Recent research provided further mechanistic insights into the delivery of the CSC to the plasma membrane by the synergistic exocyst complex and microtubules (Zhu et al. 2018a). It has been shown that fusion between the exocyst complex and the target membrane relies on an interaction between CSI1, a microtubule-binding protein that tethers the CSC to the cortical microtubules (Gu et al. 2010), and PTL1, a plant protein bearing a domain, the MUN domain, which functions in exocytosis in animal cells (Hashimoto-Sugimoto et al. 2013; Zhu et al. 2018a). Thus, this final delivery step most likely employs mostly the conserved exocytosis machinery as well as plant-specific regulators.

Questions remain about the driving force of CSC transport from Golgi and recycling from the plasma membrane. *Arabidopsis* Kinesin-4/FRA1 is required for deposition of cellulose on the cell wall (Zhong et al. 2002). FRA1 displays bona fide kinesin activity and mediates vesicular transport along cortical microtubules (Kong et al. 2015). However, another work demonstrated that Kinesin-4/FRA1 is involved in secretion of cell wall components, but it is not required for traffic of the CSC to the plasma membrane (Zhu et al. 2015). Therefore, it is possible that certain plant motor proteins exert diversified or nonexclusive functions that are still not fully appreciated. The possibility exists that a different cytoskeleton motor system carries the CSC carriers from TGN to the plasma membrane. Last but not least, the rather isolated positioning of GI-TGN from Golgi may allow the TGN to be distributed in close proximity of the target site of plasma membrane, thus not requiring cytoskeleton cable to transport the CSC (Bashline et al. 2014). Another example of cytoskeletal force-driven vesicular transport is SHORT-ROOT (SHR), a signaling protein that regulates cell division and differentiation in root (Spiegelman et al. 2018). Type-14 kinesin KinG directly interacts with SIEL, a protein that tethers SHR on the endosome, and is required for the intercellular movement of SHR (Spiegelman et al. 2018).

These reviewed studies of various cytoskeleton components and specific cargoes demonstrate the necessity of dissecting the client specificity of myosin, dynein, and kinesin motors as well as characterizing their adaptor proteins in plant cells. It is important to consider also the findings that the plant cytoskeleton may indirectly contribute to organelle transport. For example, as detailed earlier, the ER network interacts with the actin cytoskeleton via characterized stable anchors and potential adaptor-attached myosins (Stefano and Brandizzi 2018). Recent work illustrated that the plant ER is in close association with endosomes and that disruption of ER architecture and streaming largely affects the mobility and endocytic function of endosomes (Stefano et al. 2015). These results support the possibility of an alternative scenario of endosome transport in plant cells in which endosomes are mobilized by the actin cytoskeleton through tethering to the ER network.

8.4.5 Autophagosome

Autophagosomes are small endomembrane-bound compartments formed during autophagy, a conserved degradation and recycling pathway that is usually upregulated upon stress. In mammalian cells, the autophagosome moves on microtubules to reach the lysosome, while the actin cytoskeleton is also required for formation and traffic of autophagosomes (Monastyrska et al. 2009; Kruppa et al. 2016). The association of autophagosomes with microtubules involves the mammalian LC3 (microtubule-associated protein 1 light chain 3, homologs of yeast ATG8) family proteins and FYCO1 adaptor (Pankiv et al. 2010).

In plants, it has been shown that integrity of the microtubules is required for the induction of autophagosomes (Wang et al. 2015). The *Arabidopsis* homologs of ATG8 similarly interact with microtubules (Ketelaar et al. 2004). An autophagy cargo receptor Joka2 colocalizes with cytoskeleton components in plant cells, which is similar to its mammalian homolog NBR1 (Zientara-Rytter and Sirko 2014). Additionally, a recent study showed that NAP1, as well as other components of the actin-regulating SCAR/WAVE complex and Arp2/3 complex, are required for stress-induced autophagic responses (Wang et al. 2016b). This evidence suggests that the formation of the autophagic structures requires both stable cytoskeletal tracks and reorganization of the actin cytoskeleton.

8.5   Future Perspectives

8.5.1   *Potential Membrane–Cytoskeleton Interactions*

Although several proteins are emerging as motors and connectors of the plant endomembranes with the cytoskeleton (Table 8.1), several questions remain open. First, the mechanisms underlying the interactions between cytoskeleton and major endomembrane compartments are still largely unknown. For example, numerous studies suggest that abolishment of myosin XI motors significantly impairs ER dynamics; however, it is still unclear whether myosins directly connect with nuclear envelope, ER membrane, or vacuoles. Meanwhile, the exact localizations of more than a dozen identified myosin adaptors remain mysterious (Kurth et al. 2017). Moreover, it is yet unclear whether any anchoring mechanisms, other than the already identified proteins, may connect ER network, vacuoles, and other vesicular structures to actin filaments or microtubules.

Several questions also remain unanswered about the interactions between cytoskeleton system and the plant cell's unique membrane structures. For example, chloroplast movement to avoid intense light is achieved by membrane–actin interactions. The chloroplast outer envelope is bound to dynamically polymerizing short chloroplast-actin filaments by CHUP1 and other anchoring proteins (Wada and Kong 2018). It is still unclear whether myosin motors also contribute to chloroplast movement, and how the light signal coordinates dynamic actin polymerization and membrane-actin filament anchoring mechanisms. Stromules are tubular structures that protrude from plastids and function during immunity (Hanson and Hines 2018). The role of actin cytoskeleton in stromule formation is not understood. Stromules are long and branched tubular structures that frequently overlay actin filaments and the ER network (Hanson and Sattarzadeh 2011; Schattat et al. 2011). Stromule movement depends on actin filaments and myosin XI (Kwok and Hanson 2003; Natesan

Table 8.1 Plant proteins that are involved in endomembrane–cytoskeleton interactions

Endomembrane compartments	Protein/ interacting partners	Notes	References
Endoplasmic reticulum (ER)	Myosin XIk, XI1, XI2	Required for ER streaming and maintenance of ER morphology	Ueda et al. (2010)
	SYP73	ER membrane-associated actin-binding protein	Cao et al. (2016)
	NET3B	ER membrane-associated actin-binding protein	Wang and Hussey (2017)
	GhCFE1A	ER membrane-associated actin-binding protein in cotton fiber	Lv et al. (2015)
Golgi	Myosin XIk, XI1, XI2	Required for Golgi movement	Avisar et al. (2008), Peremyslov et al. (2008), Prokhnevsky et al. (2008), and Avisar et al. (2012)
	Kinesin-13A	Localized to Golgi and may contribute to Golgi positioning	Lu et al. (2005) and Wei et al. (2009)
ER-PM contact site (EPCS)	NET3C and VAP27	Form a complex with interactions between NET3C and VAP27, NET3C and F-actin, and VAP27 and microtubules	Wang et al. (2014)
trans-Golgi network (TGN)	RISAP	ROP effector, associated with myosin and F-actin	Stephan et al. (2014)
	BEN1 and HLB1	HLB1 interacts with TGN-localized BEN1 on actin cytoskeleton	Sparks et al. (2016)
	TGNap1	Binds to microtubules and interacts with TGN-localized Rab6 and YIP4	Renna et al. (2018)
Vacuole	NET4A	Associated with the tonoplast in root epidermal cells	Deeks et al. (2012)
Plasma membrane (PM)	NET1A	Associated with PM and accumulate at the plasmodesmata	Deeks et al. (2012)
	NET2A	Associated with PM in the shank of pollen tube	Deeks et al. (2012)
Endosomes and vesicles	CLASP and SNX1	Connects microtubules and endosomes that transport PM-localized proteins	Ambrose et al. (2013)
	HopW1	A bacterial effector protein that disrupts F-actin and inhibits endocytosis in plant cells	Kang et al. (2014)
	CSI1 and PTL1	Connects microtubules and delivers CSC-tethered vesicle to the PM	Gu et al. (2010) and Zhu et al. (2018a)
	Kinesin-4/ FRA1	Involved in secretion of cell wall components	Kong et al. (2015) and Zhu et al. (2015)
	KinG and SIEL	SIEL connects KinG to vesicles that transport SHR	Spiegelman et al. (2018)

(continued)

Table 8.1 (continued)

Endomembrane compartments	Protein/ interacting partners	Notes	References
Autophagosome	ATG8	Associated with microtubules	Ketelaar et al. (2004)
	NAP1	Components of SCAR/WAVE complex and Arp2/3 complex are required for formation of autophagosomes	Wang et al. (2016b)

List of plant motors that mobilize particular endomembrane compartments, as well as proteins that constitute direct connections between endomembranes and the cytoskeleton. This list includes a number of recently identified plant-specific proteins and evolutionarily conserved homologs that have specialized for plant cellular activities. These findings unveiled unappreciated mechanisms of bringing membranes to the cytoskeleton and further demonstrated the importance of membrane-cytoskeleton interactions for the plant life

et al. 2009). Besides, CHUP1, the anchor between chloroplast envelope and actin filament, is involved in stromule formation (Caplan et al. 2015). However, isolated chloroplasts without actin filaments or ATP for myosin also spontaneously form stromules (Ho and Theg 2016), raising the question on the identity of the underlying mechanism.

Little is also known about the role of the cytoskeleton in mediating organelle–organelle interactions in plant cells. Organelle–organelle interactions are widely observed and are required for organelle dynamics, collaborated metabolic processes, and signal transductions (Settembre et al. 2013; Hurlock et al. 2014; Tatsuta et al. 2014; Barbosa et al. 2015; Phillips and Voeltz 2016). Imaging cultured mammalian cells with higher resolution further illustrated that these interactions are frequent and dynamic as expected, while they can also be intense and simultaneously occur between multiple organelles (Valm et al. 2017). Studies also illustrated that mechanisms of cytoskeleton system components mediated organelle–organelle interactions, such as ER-associated mitochondrial and endosomal division (Rowland et al. 2014; Lee et al. 2016; Hoyer et al. 2018). Studies in plant cells have reported similar organelle–organelle interactions and their participation in organelle dynamics and metabolic processes (Boevink et al. 1996b, 1998; Staehelin 1997; Sparkes et al. 2011; Mehrshahi et al. 2013; Stefano et al. 2015; Gao et al. 2016). Among these topics, a convergence of the actin and microtubules at EPCS has been investigated in plant cells (Wang et al. 2017a), but further research is required to reveal the function of such an organization.

8.5.2 What Are the Upstream Inputs of Cytoskeletal Signaling?

Studies of the plant actin cytoskeleton that combined biochemical, genetic, and imaging approaches have been instrumental to appreciate the actin dynamics that are directly regulated by various actin-binding proteins (Staiger 2000; Li et al. 2015a). Several actin-binding proteins are directly regulated by ROP GTPases in plant cells, in a manner that is conserved to Rac and Rho GTPases in yeast and mammalian cells (Hall and Nobes 2000; Feiguelman et al. 2018). Besides, membrane phosphoinositides such as PI $(4,5)P_2$ and PI $(3,4,5)P_3$ might be involved in signaling to various actin-binding proteins in the cytosol, but the detailed mechanisms are not clear (Bezanilla et al. 2015). A third major regulatory mechanism for actin-binding proteins is direct phosphorylation. However, many yeast and mammalian kinases involved in this process, such as certain members of the AGC family of protein kinases, seem not existing or have adapted functions in plants (Rademacher and Offringa 2012). In addition to conventional actin-binding proteins, it would be also intriguing to explore potential regulatory mechanisms for the more recently identified plant membrane-associated actin-binding proteins, such as NETs, VAPs, and SYP73. Beyond the level of direct regulation of actin-binding proteins, certain types of receptors on the mammalian cell plasma membrane transduce extracellular signals to activate cytoplasmic actin cytoskeleton components (Brunton et al. 2004; Vicente-Manzanares et al. 2009; Vazquez-Victorio et al. 2016). To conclude, further analysis of the regulatory mechanisms of the conserved and novel actin-binding proteins has the potential to illustrate a landscape of endomembrane compartment morphogenesis and dynamics integrated by membrane–cytoskeleton interactions.

8.5.3 From Subcellular Structures to Cell Functions

One rule of thumb whereby biologists view this world is that structures determine functions and functions are fulfilled by specific structures. As our knowledge of establishment and maintenance of individual organelles is increasing in depth, intriguing questions arise as to how organelles and cellular structures such as cytoskeleton interact and collaborate to realize specific cellular functions in response to extracellular stimuli.

Several plant cell models have been adopted to study this question. Polar growth of pollen tube and root hair cells requires actomyosin system-propelled vesicular trafficking in the cell apex and actin filament-facilitated organelle extension in the elongating cell (Cai et al. 2015; Fu 2015; Qi et al. 2016; Griffing et al. 2017). Stomatal opening and closure are triggered by various signals and directly achieved through dynamic fusion of vacuoles, rearrangement of microtubule arrays and actin filaments, as well as cell wall modification (Lucas et al. 2006; Eisinger et al. 2012). Actin-dependent vacuole fusion is a general mechanism that also applies to vacuoles in guard cells (Li et al. 2013). Recent work showed that stomatal movement also requires microtubule-directed cellulose synthase complex movement on the plasma membrane (Rui and Anderson 2016).

Numerous studies revealed the reorganization of actin cytoskeleton as an important cellular response for plant defense, during both effector triggered immunity

(ETI) that is initiated from the cytoplasm and pattern-triggered immunity (PTI) that is perceived by receptors on the plasma membrane (Day et al. 2011; Li and Day 2019). Actin-binding capping proteins reside on the cytosolic side of ER membrane and mediate actin remodeling during PTI (Jimenez-Lopez et al. 2014; Li et al. 2015b). Another actin-binding protein profilin is also involved in PTI (Sun et al. 2018). Meanwhile, little is known about the signal transduction from receptors on the plasma membrane to actin-binding proteins in the cytosol, or how this process is potentially affected by the association between actin-binding proteins and endomembranes. Interestingly, recent work reported that the membrane-actin anchoring protein CHUP1 facilitates chloroplast movement toward pathogen interface (Toufexi et al. 2019), suggesting that plant defense can directly employ membrane–cytoskeleton interactions to mobilize organelles to battle pathogenesis. To conclude, further studies on these topics would illustrate how the plant cell endomembrane and cytoskeleton collaborate in response to developmental and environmental clues.

Acknowledgments We thank Dr. Sang-Jin Kim for helpful discussion. This work was primarily supported by NSF MCB1714561 and AgBioResearch MICL02598 to FB. We acknowledge infrastructure support by the Chemical Sciences, Geosciences and Biosciences Division, Office of BES, Office of Science, US DOE DE-FG02-91ER20021.

References

Akkerman M, Overdijk EJR, Schel JHN, Emons AMC, Ketelaar T (2011) Golgi body motility in the plant cell cortex correlates with actin cytoskeleton organization. Plant Cell Physiol 52:1844–1855

Ambrose C, Ruan Y, Gardiner J, Tamblyn LM, Catching A, Kirik V, Marc J, Overall R, Wasteneys GO (2013) CLASP interacts with sorting nexin 1 to link microtubules and auxin transport via PIN2 recycling in *Arabidopsis thaliana*. Dev Cell 24:649–659

Angelos E, Ruberti C, Kim SJ, Brandizzi F (2017) Maintaining the factory: the roles of the unfolded protein response in cellular homeostasis in plants. Plant J 90:671–682

Avisar D, Prokhnevsky AI, Makarova KS, Koonin EV, Dolja VV (2008) Myosin XI-K is required for rapid trafficking of Golgi stacks, peroxisomes, and mitochondria in leaf cells of *Nicotiana benthamiana*. Plant Physiol 146:1098–1108

Avisar D, Abu-Abied M, Belausov E, Sadot E (2012) Myosin XIK is a major player in cytoplasm dynamics and is regulated by two amino acids in its tail. J Exp Bot 63:241–249

Baluska F, Hlavacka A, Samaj J, Palme K, Robinson DG, Matoh T, McCurdy DW, Menzel D, Volkmann D (2002) F-actin-dependent endocytosis of cell wall pectins in meristematic root cells. Insights from brefeldin A-induced compartments. Plant Physiol 130:422–431

Barbosa AD, Savage DB, Siniossoglou S (2015) Lipid droplet-organelle interactions: emerging roles in lipid metabolism. Curr Opin Cell Biol 35:91–97

Bashline L, Li SD, Gu Y (2014) The trafficking of the cellulose synthase complex in higher plants. Ann Bot 114:1059–1067

Bayer EM, Sparkes I, Vanneste S, Rosado A (2017) From shaping organelles to signalling platforms: the emerging functions of plant ER-PM contact sites. Curr Opin Plant Biol 40:89–96

Beck M, Zhou J, Faulkner C, MacLean D, Robatzek S (2012) Spatio-temporal cellular dynamics of the Arabidopsis flagellin receptor reveal activation status-dependent endosomal sorting. Plant Cell 24:4205–4219

Bezanilla M, Gladfelter AS, Kovar DR, Lee WL (2015) Cytoskeletal dynamics: a view from the membrane. J Cell Biol 209:329–337

Bodman JAR, Yang Y, Logan MR, Eitzen G (2015) Yeast translation elongation factor-1A binds vacuole-localized Rho1p to facilitate membrane integrity through F-actin remodeling. J Biol Chem 290:4705–4716

Boevink P, SantaCruz S, Oparka KJ, Hawes C (1996a) Virus-mediated delivery of the green fluorescent protein to the endoplasmic reticulum of tobacco cells. Mol Biol Cell 7:426–426

Boevink P, SantaCruz S, Hawes C, Harris N, Oparka KJ (1996b) Virus-mediated delivery of the green fluorescent protein to the endoplasmic reticulum of plant cells. Plant J 10:935–941

Boevink P, Oparka K, Cruz SS, Martin B, Betteridge A, Hawes C (1998) Stacks on tracks: the plant Golgi apparatus traffics on an actin/ER network. Plant J 15:441–447

Bola B, Allan V (2009) How and why does the endoplasmic reticulum move? Biochem Soc Trans 37:961–965

Brandizzi F, Snapp EL, Roberts AG, Lippincott-Schwartz J, Hawes C (2002) Membrane protein transport between the endoplasmic reticulum and the Golgi in tobacco leaves is energy dependent but cytoskeleton independent: evidence from selective photobleaching. Plant Cell 14:1293–1309

Breeze E, Dzimitrowicz N, Kriechbaumer V, Brooks R, Botchway SW, Brady JP, Hawes C, Dixon AM, Schnell JR, Fricker MD, Frigerio L (2016) A C-terminal amphipathic helix is necessary for the in vivo tubule-shaping function of a plant reticulon. Proc Natl Acad Sci USA 113:10902–10907

Breuer D, Nowak J, Ivakov A, Somssich M, Persson S, Nikoloski Z (2017) System-wide organization of actin cytoskeleton determines organelle transport in hypocotyl plant cells. Proc Natl Acad Sci USA 114:E5741–E5749

Brunton VG, MacPherson IRJ, Frame MC (2004) Cell adhesion receptors, tyrosine kinases and actin modulators: a complex three-way circuitry. Biochim Biophys Acta 1692:121–144

Cai G, Parrotta L, Cresti M (2015) Organelle trafficking, the cytoskeleton, and pollen tube growth. J Integr Plant Biol 57:63–78

Cao P, Renna L, Stefano G, Brandizzi F (2016) SYP73 anchors the ER to the actin cytoskeleton for maintenance of ER integrity and streaming in Arabidopsis. Curr Biol 26:3245–3254

Caplan JL, Kumar AS, Park E, Padmanabhan MS, Hoban K, Modla S, Czymmek K, Dinesh-Kumar SP (2015) Chloroplast stromules function during innate immunity. Dev Cell 34:45–57

Chang CL, Chen YJ, Quintanilla CG, Hsieh TS, Liou J (2018) EB1 binding restricts STIM1 translocation to ER-PM junctions and regulates store-operated Ca2+ entry. J Cell Biol 217:2047–2058

Chen J, Stefano G, Brandizzi F, Zheng HQ (2011) Arabidopsis RHD3 mediates the generation of the tubular ER network and is required for Golgi distribution and motility in plant cells. J Cell Sci 124:2241–2252

Chen SL, Novick P, Ferro-Novick S (2012) ER network formation requires a balance of the dynamin-like GTPase Sey1p and the Lunapark family member Lnp1p. Nat Cell Biol 14:707–716

Crowell EF, Bischoff V, Desprez T, Rolland A, Stierhof YD, Schumacher K, Gonneau M, Hofte H, Vernhettes S (2009) Pausing of Golgi bodies on microtubules regulates secretion of cellulose synthase complexes in Arabidopsis. Plant Cell 21:1141–1154

Cui Y, Cao W, He Y, Zhao Q, Wakazaki M, Zhuang X, Gao J, Zeng Y, Gao C, Ding Y (2019) A whole-cell electron tomography model of vacuole biogenesis in Arabidopsis root cells. Nat Plants 5:95–105

Day B, Henty JL, Porter KJ, Staiger CJ (2011) The pathogen-actin connection: a platform for defense signaling in plants. Annu Rev Phytopathol 49:483–506

Deeks MJ, Calcutt JR, Ingle EK, Hawkins TJ, Chapman S, Richardson AC, Mentlak DA, Dixon MR, Cartwright F, Smertenko AP (2012) A superfamily of actin-binding proteins at the actin-membrane nexus of higher plants. Curr Biol 22:1595–1600

del Dedo JE, Idrissi F-Z, Fernandez-Golbano IM, Garcia P, Rebollo E, Krzyzanowski MK, Grötsch H, Geli MI (2017) ORP-mediated ER contact with endocytic sites facilitates actin polymerization. Dev. Cell 43:588–602.e6

Dhonukshe P, Mathur J, Hulskamp M, Gadella TWJ (2005) Microtubule plus-ends reveal essential links between intracellular polarization and localized modulation of endocytosis during division-plane establishment in plant cells. BMC Biol 3:11

Eisinger W, Ehrhardt D, Briggs W (2012) Microtubules are essential for guard-cell function in Vicia and Arabidopsis. Mol Plant 5:601–610

Eitzen G (2003) Actin remodeling to facilitate membrane fusion. Biochim Biophys Acta 1641:175–181

Eitzen G, Thorngren N, Wickner W (2001) Rho1p and Cdc42p act after Ypt7p to regulate vacuole docking. EMBO J 20:5650–5656

Eitzen G, Wang L, Thorngren N, Wickner W (2002) Remodeling of organelle-bound actin is required for yeast vacuole fusion. J Cell Biol 158:669–679

English AR, Voeltz GK (2013) Rab10 GTPase regulates ER dynamics and morphology. Nat Cell Biol 15:169–178

Feeney M, Kittelmann M, Menassa R, Hawes C, Frigerio L (2018) Protein storage vacuoles originate from remodeled preexisting vacuoles in *Arabidopsis thaliana*. Plant Physiol 177:241–254

Fehrenbacher KL, Davis D, Wu M, Boldogh I, Pon LA (2002) Endoplasmic reticulum dynamics, inheritance, and cytoskeletal interactions in budding yeast. Mol Biol Cell 13:854–865

Feiguelman G, Fu Y, Yalovsky S (2018) ROP GTPases structure-function and signaling pathways. Plant Physiol 176:57–79

Friml J (2010) Subcellular trafficking of PIN auxin efflux carriers in auxin transport. Eur J Cell Biol 89:231–235

Friml J, Wisniewska J, Benkova E, Mendgen K, Palme K (2002) Lateral relocation of auxin efflux regulator PIN3 mediates tropism in Arabidopsis. Nature 415:806–809

Fu Y (2015) The cytoskeleton in the pollen tube. Curr Opin Plant Biol 28:111–119

Gao HB, Metz J, Teanby NA, Ward AD, Botchway SW, Coles B, Pollard MR, Sparkes I (2016) In vivo quantification of peroxisome tethering to chloroplasts in tobacco epidermal cells using optical tweezers. Plant Physiol 170:263–272

Geldner N, Friml J, Stierhof YD, Jurgens G, Palme K (2001) Auxin transport inhibitors block PIN1 cycling and vesicle trafficking. Nature 413:425–428

Gosavi P, Gleeson PA (2017) The function of the Golgi ribbon structure—an enduring mystery unfolds! BioEssays 39. https://doi.org/10.1002/bies.201700063

Granger E, McNee G, Allan V, Woodman P (2014) The role of the cytoskeleton and molecular motors in endosomal dynamics. Semin Cell Dev Biol 31:20–29

Griffing LR, Gao HBT, Sparkes I (2014) ER network dynamics are differentially controlled by myosins XI-K, XI-C, XI-E, XI-I, XI-1, and XI-2. Front Plant Sci 5:218

Griffing LR, Lin C, Perico C, White RR, Sparkes I (2017) Plant ER geometry and dynamics: biophysical and cytoskeletal control during growth and biotic response. Protoplasma 254:43–56

Gu Y, Kaplinsky N, Bringmann M, Cobb A, Carroll A, Sampathkumar A, Baskin TI, Persson S, Somerville CR (2010) Identification of a cellulose synthase-associated protein required for cellulose biosynthesis. Proc Natl Acad Sci USA 107:12866–12871

Guo YT, Li D, Zhang SW, Yang YR, Liu JJ, Wang XY, Liu C, Milkie DE, Moore RP, Tulu US, Kiehart DP, Hu JJ, Lippincott-Schwartz J, Betzig E, Li D (2018) Visualizing intracellular organelle and cytoskeletal interactions at nanoscale resolution on millisecond timescales. Cell 175:1430–1442

Gutierrez R, Lindeboom JJ, Paredez AR, Emons AMC, Ehrhardt DW (2009) Arabidopsis cortical microtubules position cellulose synthase delivery to the plasma membrane and interact with cellulose synthase trafficking compartments. Nat Cell Biol 11:797–U743

Hall A, Nobes CD (2000) Rho GTPases: molecular switches that control the organization and dynamics of the actin cytoskeleton. Philos Trans R Soc Lond Ser B Biol Sci 355:965–970

Hamada T, Ueda H, Kawase T, Hara-Nishimura I (2014) Microtubules contribute to tubule elongation and anchoring of endoplasmic reticulum, resulting in high network complexity in Arabidopsis. Plant Physiol 166:1869–U1042

Hanson MR, Hines KM (2018) Stromules: probing formation and function. Plant Physiol 176:128–137

Hanson MR, Sattarzadeh A (2011) Stromules: recent insights into a long neglected feature of plastid morphology and function. Plant Physiol 155:1486–1492

Hashimoto-Sugimoto M, Higaki T, Yaeno T, Nagami A, Irie M, Fujimi M, Miyamoto M, Akita K, Negi J, Shirasu K, Hasezawa S, Iba K (2013) A Munc13-like protein in Arabidopsis mediates H +-ATPase translocation that is essential for stomatal responses. Nat Commun 4:2215

Henne WM, Liou J, Emr SD (2015) Molecular mechanisms of inter-organelle ER-PM contact sites. Curr Opin Cell Biol 35:123–130

Higaki T, Kutsuna N, Okubo E, Sano T, Hasezawa S (2006) Actin microfilaments regulate vacuolar structures and dynamics: dual observation of actin microfilaments and vacuolar membrane in living tobacco BY-2 cells. Plant Cell Physiol 47:839–852

Ho J, Theg SM (2016) The formation of stromules in vitro from chloroplasts isolated from Nicotiana benthamiana. PLoS One 11:1–14

Hoffmann A, Nebenführ A (2004) Dynamic rearrangements of transvacuolar strands in BY-2 cells imply a role of myosin in remodeling the plant actin cytoskeleton. Protoplasma 224:201–210

Hoyer MJ, Chitwood PJ, Ebmeier CC, Striepen JF, Qi RZ, Old WM, Voeltz GK (2018) A novel class of ER membrane proteins regulates ER-associated endosome fission. Cell 175:254–265

Hu JJ, Shibata Y, Zhu PP, Voss C, Rismanchi N, Prinz WA, Rapoport TA, Blackstone C (2009) A class of dynamin-like GTPases involved in the generation of the tubular ER network. Cell 138:549–561

Hu YF, Na XF, Li JL, Yang LJ, You J, Liang XL, Wang JF, Peng L, Bi YR (2015) Narciclasine, a potential allelochemical, affects subcellular trafficking of auxin transporter proteins and actin cytoskeleton dynamics in Arabidopsis roots. Planta 242:1349–1360

Hurlock AK, Roston RL, Wang K, Benning C (2014) Lipid trafficking in plant cells. Traffic 15:915–932

Isgandarova S, Jones L, Forsberg D, Loncar A, Dawson J, Tedrick K, Eitzen G (2007) Stimulation of actin polymerization by vacuoles via Cdc42p-dependent signaling. J Biol Chem 282:30466–30475

Ivanov R, Brumbarova T, Blum A, Jantke AM, Fink-Straube C, Bauer P (2014) SORTING NEXIN1 is required for modulating the trafficking and stability of the Arabidopsis IRON-REGULATED TRANSPORTER1. Plant Cell 26:1294–1307

Jimenez-Lopez JC, Wang X, Kotchoni SO, Huang SJ, Szymanski DB, Staiger CJ (2014) Heterodimeric capping protein from Arabidopsis is a membrane-associated, actin-binding protein. Plant Physiol 166:1312–1328

Joensuu M, Belevich I, Ramo O, Nevzorov I, Vihinen H, Puhka M, Witkos TM, Lowe M, Vartiainen MK, Jokitalo E (2014) ER sheet persistence is coupled to myosin 1c-regulated dynamic actin filament arrays. Mol Biol Cell 25:1111–1126

Jones L, Tedrick K, Baier A, Logan MR, Eitzen G (2010) Cdc42p is activated during vacuole membrane fusion in a sterol-dependent subreaction of priming. J Biol Chem 285:4298–4306

Kang YS, Jelenska J, Cecchini NM, Li YJ, Lee MW, Kovar DR, Greenberg JT (2014) HopW1 from Pseudomonas syringae disrupts the actin cytoskeleton to promote virulence in Arabidopsis. PLoS Path 10:1–10

Ketelaar T, Voss C, Dimmock SA, Thumm M, Hussey PJ (2004) Arabidopsis homologues of the autophagy protein Atg8 are a novel family of microtubule binding proteins. FEBS Lett 567:302–306

Kim H, Park M, Kim SJ, Hwang I (2005) Actin filaments play a critical role in vacuolar trafficking at the Golgi complex in plant cells. Plant Cell 17:888–902

Kleine-Vehn J, Dhonukshe P, Swarup R, Bennett M, Friml J (2006) Subcellular trafficking of the Arabidopsis auxin influx carrier AUX1 uses a novel pathway distinct from PIN1. Plant Cell 18:3171–3181

Kong ZS, Ioki M, Braybrook S, Li SD, Ye ZH, Lee YRJ, Hotta T, Chang A, Tian J, Wang GD, Liu B (2015) Kinesin-4 functions in vesicular transport on cortical microtubules and regulates cell wall mechanics during cell elongation in plants. Mol Plant 8:1011–1023

Kriechbaumer V, Botchway SW, Slade SE, Knox K, Frigerio L, Oparka K, Hawes C (2015) Reticulomics: protein-protein interaction studies with two plasmodesmata-localized reticulon family proteins identify binding partners enriched at plasmodesmata, endoplasmic reticulum, and the plasma membrane. Plant Physiol 169:1933–1945

Kriechbaumer V, Breeze E, Pain C, Tolmie F, Frigerio L, Hawes C (2018) Arabidopsis Lunapark proteins are involved in ER cisternae formation. New Phytol 219:990–1004

Kruppa AJ, Kendrick-Jones J, Buss F (2016) Myosins, actin and autophagy. Traffic 17:878–890

Kurth EG, Peremyslov VV, Turner HL, Makarova KS, Iranzo J, Mekhedov SL, Koonin EV, Dolja VV (2017) Myosin-driven transport network in plants. Proc Natl Acad Sci USA 114:E1385–E1394

Kutsuna N, Kumagai F, Sato MH, Hasezawa S (2003) Three-dimensional reconstruction of tubular structure of vacuolar membrane throughout mitosis in living tobacco cells. Plant Cell Physiol 44:1045–1054

Kwok EY, Hanson MR (2003) Microfilaments and microtubules control the morphology and movement of non-green plastids and stromules in *Nicotiana tabacum*. Plant J 35:16–26

Lee H, Sparkes I, Gattolin S, Dzimitrowicz N, Roberts LM, Hawes C, Frigerio L (2013) An Arabidopsis reticulon and the atlastin homologue RHD3-like2 act together in shaping the tubular endoplasmic reticulum. New Phytol 197:481–489

Lee JE, Westrate LM, Wu HX, Page C, Voeltz GK (2016) Multiple dynamin family members collaborate to drive mitochondrial division. Nature 540:139–143

Lee E, Vanneste S, Perez-Sancho J, Benitez-Fuente F, Strelau M, Macho AP, Botella MA, Friml J, Rosado A (2019) Ionic stress enhances ER-PM connectivity via phosphoinositide-associated SYT1 contact site expansion in Arabidopsis. Proc Natl Acad Sci USA 116:1420–1429

Lewis JD, Lazarowitz SG (2010) Arabidopsis synaptotagmin SYTA regulates endocytosis and virus movement protein cell-to-cell transport. Proc Natl Acad Sci USA 107:2491–2496

Li P, Day B (2019) Battlefield cytoskeleton: turning the tide on plant immunity. Mol Plant-Microbe Interact 32:25–34

Li SC, Kane PM (2009) The yeast lysosome-like vacuole: endpoint and crossroads. Biochim Biophys Acta 1793:650–663

Li LJ, Ren F, Gao XQ, Wei PC, Wang XC (2013) The reorganization of actin filaments is required for vacuolar fusion of guard cells during stomatal opening in Arabidopsis. Plant Cell Environ 36:484–497

Li JJ, Blanchoin L, Staiger CJ (2015a) Signaling to actin stochastic dynamics. Annu Rev Plant Biol 66:415–440

Li JJ, Henty-Ridilla JL, Staiger BH, Day B, Staiger CJ (2015b) Capping protein integrates multiple MAMP signalling pathways to modulate actin dynamics during plant innate immunity. Nat Commun 6:7206

Löfke C, Dünser K, Scheuring D, Kleine-Vehn J (2015) Auxin regulates SNARE-dependent vacuolar morphology restricting cell size. elife 4:e05868

Lu L, Lee YRJ, Pan RQ, Maloof JN, Liu B (2005) An internal motor kinesin is associated with the Golgi apparatus and plays a role in trichome morphogenesis in Arabidopsis. Mol Biol Cell 16:811–823

Lucas JR, Nadeau JA, Sack FD (2006) Microtubule arrays and Arabidopsis stomatal development. J Exp Bot 57:71–79

Lv FN, Wang HH, Wang XY, Han LB, Ma YP, Wang S, Feng ZD, Niu XW, Cai CP, Kong ZS, Zhang TZ, Guo WZ (2015) GhCFE1A, a dynamic linker between the ER network and actin cytoskeleton, plays an important role in cotton fibre cell initiation and elongation. J Exp Bot 66:1877–1889

Ma B, Qian D, Nan Q, Tan C, An L, Xiang Y (2012) Arabidopsis vacuolar H+-ATPase (V-ATPase) B subunits are involved in actin cytoskeleton remodeling via binding to, bundling, and stabilizing F-actin. J Biol Chem 287:19008–19017

Madison SL, Buchanan ML, Glass JD, McClain TF, Park E, Nebenfuhr A (2015) Class XI myosins move specific organelles in pollen tubes and are required for normal fertility and pollen tube growth in Arabidopsis. Plant Physiol 169:1946–1960

Mathur J, Mathur N, Kernebeck B, Srinivas BP, Hulskamp M (2003) A novel localization pattern for an EB1-like protein links microtubule dynamics to endomembrane organization. Curr Biol 13:1991–1997

Meagher RB, Fechheimer M (2003) The Arabidopsis cytoskeletal genome. Arabidopsis Book 2: e0096

Mehrshahi P, Stefano G, Andaloro JM, Brandizzi F, Froehlich JE, DellaPenna D (2013) Transorganellar complementation redefines the biochemical continuity of endoplasmic reticulum and chloroplasts. Proc Natl Acad Sci USA 110:12126–12131

Molines AT, Marion J, Chabout S, Besse L, Dompierre JP, Mouille G, Coquelle FM (2018) EB1 contributes to microtubule bundling and organization, along with root growth, in Arabidopsis thaliana. Biol Open 7:bio030510

Monastyrska I, Rieter E, Klionsky DJ, Reggiori F (2009) Multiple roles of the cytoskeleton in autophagy. Biol Rev 84:431–448

Muller O, Johnson DI, Mayer A (2001) Cdc42p functions at the docking stage of yeast vacuole membrane fusion. EMBO J 20:5657–5665

Natesan SKA, Sullivan JA, Gray JC (2009) Myosin XI is required for actin-associated movement of plastid stromules. Mol Plant 2:1262–1272

Nebenfuhr A, Dixit R (2018) Kinesins and myosins: molecular motors that coordinate cellular functions in plants. Annu Rev Plant Biol 69:329–361

Nebenfuhr A, Gallagher LA, Dunahay TG, Frohlick JA, Mazurkiewicz AM, Meehl JB, Staehelin LA (1999) Stop-and-go movements of plant Golgi stacks are mediated by the acto-myosin system. Plant Physiol 121:1127–1141

Niemes S, Labs M, Scheuring D, Krueger F, Langhans M, Jesenofsky B, Robinson DG, Pimpl P (2010) Sorting of plant vacuolar proteins is initiated in the ER. Plant J 62:601–614

Nixon-Abell J, Obara CJ, Weigel AV, Li D, Legant WR, Xu C, Pasolli H, Harvey K, Hess HF, Betzig E, Blackstone CD, Lippincott-Schwartz J (2016) Increased spatiotemporal resolution reveals highly dynamic dense tubular matrices in the peripheral ER. Science 354:aaf3928

Pankiv S, Alemu EA, Brech A, Bruun JA, Lamark T, Overvatn A, Bjorkoy G, Johansen T (2010) FYCO1 is a Rab7 effector that binds to LC3 and PI3P to mediate microtubule plus end-directed vesicle transport. J Cell Biol 188:253–269

Peremyslov VV, Prokhnevsky AI, Avisar D, Dolja VV (2008) Two class XI myosins function in organelle trafficking and root hair development in Arabidopsis. Plant Physiol 146:1109–1116

Peremyslov VV, Prokhnevsky AI, Dolja VV (2010) Class XI myosins are required for development, cell expansion, and F-actin organization in Arabidopsis. Plant Cell 22:1883–1897

Peremyslov VV, Klocko AL, Fowler JE, Dolja VV (2012) Arabidopsis myosin XI-K localizes to the motile endomembrane vesicles associated with F-actin. Front Plant Sci 3:184

Peremyslov VV, Morgun EA, Kurth EG, Makarova KS, Koonin EV, Dolja VV (2013) Identification of myosin XI receptors in Arabidopsis defines a distinct class of transport vesicles. Plant Cell 25:3022–3038

Phillips MJ, Voeltz GK (2016) Structure and function of ER membrane contact sites with other organelles. Nat Rev Mol Cell Biol 17:69–82

Pollard TD, Goldman RD (2018) Overview of the cytoskeleton from an evolutionary perspective. Cold Spring Harb Perspect Biol 10:a030288

Poteryaev D, Squirrell JM, Campbell JM, White JG, Spang A (2005) Involvement of the actin cytoskeleton and homotypic membrane fusion in ER dynamics in *Caenorhabditis elegans*. Mol Biol Cell 16:2139–2153

Pourcher M, Santambrogio M, Thazar N, Thierry AM, Fobis-Loisy I, Miege C, Jaillais Y, Gaude T (2010) Analyses of SORTING NEXINs reveal distinct retromer-subcomplex functions in development and protein sorting in *Arabidopsis thaliana*. Plant Cell 22:3980–3991

Powers RE, Wang SY, Liu TY, Rapoport TA (2017) Reconstitution of the tubular endoplasmic reticulum network with purified components. Nature 543:257–260

Prinz WA, Grzyb L, Veenhuis M, Kahana JA, Silver PA, Rapoport TA (2000) Mutants affecting the structure of the cortical endoplasmic reticulum in *Saccharomyces cerevisiae*. J Cell Biol 150:461–474

Prokhnevsky AI, Peremyslov VV, Dolja VV (2008) Overlapping functions of the four class XI myosins in Arabidopsis growth, root hair elongation, and organelle motility. Proc Natl Acad Sci USA 105:19744–19749

Qi XY, Sun JQ, Zheng HQ (2016) A GTPase-dependent fine ER is required for localized secretion in polarized growth of root hairs. Plant Physiol 171:1996–2007

Quader H (1990) Formation and disintegration of cisternae of the endoplasmic reticulum visualized in live cells by conventional fluorescence and confocal laser scanning microscopy: evidence for the involvement of calcium and the cytoskeleton. Protoplasma 155:166–175

Quader H, Hofmann A, Schnepf E (1989) Reorganization of the endoplasmic reticulum in epidermal cells of onion bulb scales after cold stress—involvement of cytoskeletal elements. Planta 177:273–280

Rademacher EH, Offringa R (2012) Evolutionary adaptations of plant AGC kinases: from light signaling to cell polarity regulation. Front Plant Sci 3:250

Rakusová H, Han H, Valošek P, Friml J (2019) Genetic screen for factors mediating PIN polarization in gravistimulated *Arabidopsis thaliana* hypocotyls. Plant J 98:1048–1059

Renna L, Stefano G, Slabaugh E, Wormsbaecher C, Sulpizio A, Zienkiewicz K, Brandizzi F (2018) TGNap1 is required for microtubule-dependent homeostasis of a subpopulation of the plant trans-Golgi network. Nat Commun 9:5313

Rowland AA, Chitwood PJ, Phillips MJ, Voeltz GK (2014) ER contact sites define the position and timing of endosome fission. Cell 159:1027–1041

Rui Y, Anderson CT (2016) Functional analysis of cellulose and xyloglucan in the walls of stomatal guard cells of Arabidopsis. Plant Physiol 170:1398–1419

Saravanan RS, Slabaugh E, Singh VR, Lapidus LJ, Haas T, Brandizzi F (2009) The targeting of the oxysterol-binding protein ORP3a to the endoplasmic reticulum relies on the plant VAP33 homolog PVA12. Plant J 58:817–830

Schattat M, Barton K, Baudisch B, Klosgen RB, Mathur J (2011) Plastid stromule branching coincides with contiguous endoplasmic reticulum dynamics. Plant Physiol 155:1667–1677

Scheuring D, Löfke C, Krüger F, Kittelmann M, Eisa A, Hughes L, Smith RS, Hawes C, Schumacher K, Kleine-Vehn J (2016) Actin-dependent vacuolar occupancy of the cell determines auxin-induced growth repression. Proc Natl Acad Sci USA 113:452–457

Settembre C, Fraldi A, Medina DL, Ballabio A (2013) Signals from the lysosome: a control centre for cellular clearance and energy metabolism. Nat Rev Mol Cell Biol 14:283–296

Sheahan MB, Rose RJ, McCurdy DW (2007) Actin-filament-dependent remodeling of the vacuole in cultured mesophyll protoplasts. Protoplasma 230:141–152

Shemesh T, Klemm RW, Romano FB, Wang SY, Vaughan J, Zhuang XW, Tukachinsky H, Kozlov MM, Rapoport TA (2014) A model for the generation and interconversion of ER morphologies. Proc Natl Acad Sci USA 111:E5243–E5251

Shibata Y, Voss C, Rist JM, Hu J, Rapoport TA, Prinz WA, Voeltz GK (2008) The reticulon and DP1/Yop1p proteins form immobile oligomers in the tubular endoplasmic reticulum. J Biol Chem 283:18892–18904

Shibata Y, Hu JJ, Kozlov MM, Rapoport TA (2009) Mechanisms shaping the membranes of cellular organelles. Annu Rev Cell Dev Biol 25:329–354

Shibata Y, Shemesh T, Prinz WA, Palazzo AF, Kozlov MM, Rapoport TA (2010) Mechanisms determining the morphology of the peripheral ER. Cell 143:774–788

Siao W, Wang PW, Voigt B, Hussey PJ, Baluska F (2016) Arabidopsis SYT1 maintains stability of cortical endoplasmic reticulum networks and VAP27-1-enriched endoplasmic reticulum-plasma membrane contact sites. J Exp Bot 67:6161–6171

Simonetti B, Cullen PJ (2019) Actin-dependent endosomal receptor recycling. Curr Opin Cell Biol 56:22–33

Sparkes I, Runions J, Hawes C, Griffing L (2009a) Movement and remodeling of the endoplasmic reticulum in nondividing cells of tobacco leaves. Plant Cell 21:3937–3949

Sparkes IA, Ketelaar T, de Ruijter NCA, Hawes C (2009b) Grab a Golgi: laser trapping of Golgi bodies reveals in vivo interactions with the endoplasmic reticulum. Traffic 10:567–571

Sparkes I, Tolley N, Aller I, Svozil J, Osterrieder A, Botchway S, Mueller C, Frigerio L, Hawes C (2010) Five Arabidopsis reticulon isoforms share endoplasmic reticulum location, topology, and membrane-shaping properties. Plant Cell 22:1333–1343

Sparkes IA, Graumann K, Martiniere A, Schoberer J, Wang P, Osterrieder A (2011) Bleach it, switch it, bounce it, pull it: using lasers to reveal plant cell dynamics. J Exp Bot 62:1–7

Sparks JA, Kwon T, Renna L, Liao FQ, Brandizzi F, Blancaflor EB (2016) HLB1 is a tetratricopeptide repeat domain-containing protein that operates at the intersection of the exocytic and endocytic pathways at the TGN/EE in Arabidopsis. Plant Cell 28:746–769

Spiegelman Z, Lee CM, Gallagher KL (2018) KinG is a plant-specific kinesin that regulates both intra- and intercellular movement of SHORT-ROOT. Plant Physiol 176:392–405

Staehelin LA (1997) The plant ER: a dynamic organelle composed of a large number of discrete functional domains. Plant J 11:1151–1165

Staiger CJ (2000) Signaling to the actin cytoskeleton in plants. Annu Rev Plant Physiol Plant Mol Biol 51:257–288

Staiger CJ, Yuan M, Valenta R, Shaw PJ, Warn RM, Lloyd CW (1994) Microinjected profilin affects cytoplasmic streaming in plant cells by rapidly depolymerizing actin microfilaments. Curr Biol 4:215–219

Stefano G, Brandizzi F (2018) Advances in plant ER architecture and dynamics. Plant Physiol 176:178–186

Stefano G, Renna L, Moss T, Mcnew JA, Brandizzi F (2012) In Arabidopsis, the spatial and dynamic organization of the endoplasmic reticulum and Golgi apparatus is influenced by the integrity of the C-terminal domain of RHD3, a non-essential GTPase. Plant J 69:957–966

Stefano G, Hawes C, Brandizzi F (2014a) ER—the key to the highway. Curr Opin Plant Biol 22:30–38

Stefano G, Renna L, Brandizzi F (2014b) The endoplasmic reticulum exerts control over organelle streaming during cell expansion. J Cell Sci 127:947–953

Stefano G, Renna L, Lai YS, Slabaugh E, Mannino N, Buono RA, Otegui MS, Brandizzi F (2015) ER network homeostasis is critical for plant endosome streaming and endocytosis. Cell Discov 1:15033

Stefano G, Renna L, Wormsbaecher C, Gamble J, Zienkiewicz K, Brandizzi F (2018) Plant endocytosis requires the ER membrane-anchored proteins VAP27-1 and VAP27-3. Cell Rep 23:2299–2307

Stephan O, Cottier S, Fahlen S, Montes-Rodriguez A, Sun J, Eklund DM, Klahre U, Kost B (2014) RISAP is a TGN-associated RAC5 effector regulating membrane traffic during polar cell growth in tobacco. Plant Cell 26:4426–4447

Sun H, Qiao Z, Chua KP, Tursic A, Liu X, Gao YG, Mu YG, Hou XL, Miao YS (2018) Profilin negatively regulates formin-mediated actin assembly to modulate PAMP-triggered plant immunity. Curr Biol 28:1882–1895

Szymanski DB, Cosgrove DJ (2009) Dynamic coordination of cytoskeletal and cell wall systems during plant cell morphogenesis. Curr Biol 19:R800–R811

Taiz L (1984) Plant cell expansion: regulation of cell wall mechanical properties. Annu Rev Plant Physiol Plant Mol Biol 35:585–657

Takenawa T, Suetsugu S (2007) The WASP-WAVE protein network: connecting the membrane to the cytoskeleton. Nat Rev Mol Cell Biol 8:37–48

Tatsuta T, Scharwey M, Langer T (2014) Mitochondrial lipid trafficking. Trends Cell Biol 24:44–52

Terasaki M, Chen LB, Fujiwara K (1986) Microtubules and the endoplasmic-reticulum are highly interdependent structures. J Cell Biol 103:1557–1568

Tolley N, Sparkes IA, Hunter PR, Craddock CP, Nuttall J, Roberts LM, Hawes C, Pedrazzini E, Frigerio L (2008) Overexpression of a plant reticulon remodels the lumen of the cortical endoplasmic reticulum but does not perturb protein transport. Traffic 9:94–102

Toufexi A, Duggan C, Pandey P, Savage Z, Segretin ME, Yuen LH, Gaboriau DCA, Leary AY, Khandare V, Ward AD, Botchway SW, Bateman BC, Pan I, Schattat M, Sparkes I, Bozkurt TO (2019) Chloroplasts navigate towards the pathogen interface to counteract infection by the Irish potato famine pathogen. bioRxiv. https://doi.org/10.1101/516443

Ueda H, Yokota E, Kutsuna N, Shimada T, Tamura K, Shimmen T, Hasezawa S, Dolja VV, Hara-Nishimura I (2010) Myosin-dependent endoplasmic reticulum motility and F-actin organization in plant cells. Proc Natl Acad Sci USA 107:6894–6899

Ueda H, Tamura K, Hara-Nishimura I (2015) Functions of plant-specific myosin XI: from intracellular motility to plant postures. Curr Opin Plant Biol 28:30–38

Ueda H, Ohta N, Kimori Y, Uchida T, Shimada T, Tamura K, Hara-Nishimura I (2018) Endoplasmic reticulum (ER) membrane proteins (LUNAPARKs) are required for proper configuration of the cortical ER network in plant cells. Plant Cell Physiol 59:1931–1941

Uemura T, Suda Y, Ueda T, Nakano A (2014) Dynamic behavior of the trans-Golgi network in root tissues of Arabidopsis revealed by super-resolution live imaging. Plant Cell Physiol 55:694–703

Valm AM, Cohen S, Legant WR, Melunis J, Hershberg U, Wait E, Cohen AR, Davidson MW, Betzig E, Lippincott-Schwartz J (2017) Applying systems-level spectral imaging and analysis to reveal the organelle interactome. Nature 546:162–167

van Vliet AR, Giordano F, Gerlo S, Segura I, Van Eygen S, Molenberghs G, Rocha S, Houcine A, Derua R, Verfaillie T (2017) The ER stress sensor PERK coordinates ER-plasma membrane contact site formation through interaction with filamin-A and F-actin remodeling. Mol Cell 65:885–899.e6

Vazquez-Victorio G, Gonzalez-Espinosa C, Espinosa-Riquer ZP, Macias-Silva M (2016) GPCRs and actin–cytoskeleton dynamics. In: Shukla AK (ed) G protein-coupled receptors. Academic, London, pp 165–188

Vicente-Manzanares M, Choi CK, Horwitz AR (2009) Integrins in cell migration—the actin connection. J Cell Sci 122:199–206

Voeltz GK, Prinz WA, Shibata Y, Rist JM, Rapoport TA (2006) A class of membrane proteins shaping the tubular endoplasmic reticulum. Cell 124:573–586

Wada M, Kong S-G (2018) Actin-mediated movement of chloroplasts. J Cell Sci 131:jcs210310

Wang P, Hussey PJ (2017) NETWORKED 3B: a novel protein in the actin cytoskeleton-endoplasmic reticulum interaction. J Exp Bot 68:1441–1450

Wang PW, Hawkins TJ, Richardson C, Cummins I, Deeks MJ, Sparkes I, Hawes C, Hussey PJ (2014) The plant cytoskeleton, NET3C, and VAP27 mediate the link between the plasma membrane and endoplasmic reticulum. Curr Biol 24:1397–1405

Wang Y, Zheng XY, Yu BJ, Han SJ, Guo JB, Tang HP, Yu AYL, Deng HT, Hong YG, Liu YL (2015) Disruption of microtubules in plants suppresses macroautophagy and triggers starch excess-associated chloroplast autophagy. Autophagy 11:2259–2274

Wang P, Richardson C, Hawkins TJ, Sparkes I, Hawes C, Hussey PJ (2016a) Plant VAP27 proteins: domain characterization, intracellular localization and role in plant development. New Phytol 210:1311–1326

Wang PW, Richardson C, Hawes C, Hussey PJ (2016b) Arabidopsis NAP1 regulates the formation of autophagosomes. Curr Biol 26:2060–2069

Wang S, Tukachinsky H, Romano FB, Rapoport TA (2016c) Cooperation of the ER-shaping proteins atlastin, lunapark, and reticulons to generate a tubular membrane network. elife 5:e18605

Wang PW, Hawes C, Hussey PJ (2017a) Plant endoplasmic reticulum-plasma membrane contact sites. Trends Plant Sci 22:289–297

Wang X, Li S, Wang H, Shui W, Hu J (2017b) Quantitative proteomics reveal proteins enriched in tubular endoplasmic reticulum of *Saccharomyces cerevisiae*. elife 6:e23816

Waterman-Storer CM, Salmon ED (1998) Endoplasmic reticulum membrane tubules are distributed by microtubules in living cells using three distinct mechanisms. Curr Biol 8:798–806

Wei LQ, Zhang W, Liu ZH, Li Y (2009) AtKinesin-13A is located on Golgi-associated vesicle and involved in vesicle formation/budding in Arabidopsis root-cap peripheral cells. BMC Plant Biol 9:138

Westrate LM, Lee JE, Prinz WA, Voeltz GK (2015) Form follows function: the importance of endoplasmic reticulum shape. Annu Rev Biochem 84:791–811

Wickner W (2010) Membrane fusion: five lipids, four SNAREs, three chaperones, two nucleotides, and a Rab, all dancing in a ring on yeast vacuoles. Annu Rev Cell Dev Biol 26:115–136

Yadav S, Linstedt AD (2011) Golgi positioning. Cold Spring Harb Perspect Biol 3:a005322

Yang XC, Bassham DC (2015) New insight into the mechanism and function of autophagy in plant cells. In: Jeon KW (ed) International review of cell and molecular biology. Academic, Burlington, MA, pp 1–40

Yokota E, Ueda H, Hashimoto K, Orii H, Shimada T, Hara-Nishimura I, Shimmen T (2011) Myosin XI-dependent formation of tubular structures from endoplasmic reticulum isolated from tobacco cultured BY-2 cells. Plant Physiol 156:129–143

Zhang M, Wu FY, Shi JM, Zhu YM, Zhu ZM, Gong QQ, Hu JJ (2013) ROOT HAIR DEFECTIVE3 family of dynamin-like GTPases mediates homotypic endoplasmic reticulum fusion and is essential for Arabidopsis development. Plant Physiol 163:713–720

Zhang C, Hicks GR, Raikhel NV (2014) Plant vacuole morphology and vacuolar trafficking. Front Plant Sci 5:476

Zheng JM, Han SW, Rodriguez-Welsh MF, Rojas-Pierce M (2014) Homotypic vacuole fusion requires VTI11 and is regulated by phosphoinositides. Mol Plant 7:1026–1040

Zhong RQ, Burk DH, Morrison WH, Ye ZH (2002) A kinesin-like protein is essential for oriented deposition of cellulose microfibrils and cell wall strength. Plant Cell 14:3101–3117

Zhou X, He Y, Huang X, Guo Y, Li D, Hu J (2019) Reciprocal regulation between lunapark and atlastin facilitates ER three-way junction formation. Protein Cell 10:510–525

Zhu CM, Ganguly A, Baskin TI, McClosky DD, Anderson CT, Foster C, Meunier KA, Okamoto R, Berg H, Dixit R (2015) The fragile Fiber1 kinesin contributes to cortical microtubule-mediated trafficking of cell wall components. Plant Physiol 167:780–792

Zhu X, Li S, Pan S, Xin X, Gu Y (2018a) CSI1, PATROL1, and exocyst complex cooperate in delivery of cellulose synthase complexes to the plasma membrane. Proc Natl Acad Sci USA 115:E5635–E5635

Zhu YM, Zhang GM, Lin SY, Shi JM, Zhang H, Hu JJ (2018b) Sec61 beta facilitates the maintenance of endoplasmic reticulum homeostasis by associating microtubules. Protein Cell 9:616–628

Zientara-Rytter K, Sirko A (2014) Selective autophagy receptor Joka2 co-localizes with cytoskeleton in plant cells. Plant Signal Behav 9:e28523

Printed in the United States
By Bookmasters